Houda Berrouk
Saloua Ben Adouda
Nadia Chaib

Antifungal power of Artemisia herba alba and Thymus vulgaris

Houda Berrouk
Saloua Ben Adouda
Nadia Chaib

Antifungal power of Artemisia herba alba and Thymus vulgaris

Vis-à-vis some human pathogenic fungi

ScienciaScripts

Cover image: www.ingimage.com

This book is a translation from the original published under ISBN 978-620-6-69811-1.

Publisher:
Sciencia Scripts
is a trademark of
Dodo Books Indian Ocean Ltd. and OmniScriptum S.R.L publishing group

120 High Road, East Finchley, London, N2 9ED, United Kingdom
Str. Armeneasca 28/1, office 1, Chisinau MD-2012, Republic of Moldova, Europe
Printed at: see last page
ISBN: 978-620-7-66467-2

Contents

INTRODUCTION

In the past, medicinal plants were one of the only sources of cures for diseases. The knowledge that man possesses in this part of botany has not been acquired without danger. Man regards plants as faithful companions, to whom he turns for the goods they provide and for their great ийШё. They are vëritaЬles natural pharmacies, which our good god has crëë and ё1.аЬН on the earth in order to maintain our health, prevent our ills, and even cure them. [2]

The use of plants in medicine goes back a very long way. It has even been discovered that wild animals instinctively use certain plants to treat themselves! Today, if traditional medicine is to bear fruit on a large scale, and be even more effective, it needs to meet up with so-called "modern" medicine.

In addition to the fact that they constitute an indubitable biological wealth, phylogenetic resources are the source of numerous virtues thanks to the active principles they дё^тет

Some plant species, recognised as medicinal plants, are capable of producing natural substances whose specificity is closely related to the nature of the species in question [9].

In traditional medicine, the drug is used as a sëdative Ьдёге for cases of nervousness, fatigue and sleep disorders, just as it is used in the composition of many sëdative herbal teas [16].

Today, despite the development of chemical therapeutics, the use of medicinal plants is still very popular, due to their effectiveness in various therapeutic procedures, thanks to the active ingredients they contain. Among these active ingredients are essential oils, which are aromatic and volatile extracts. [24]

Essential oils are volatile, refringent, optically active liquids, similar to oils, with a distinctive odour. They are formed in a large number of plants as by-products of secondary metabolism.

Chemically speaking, they are extremely complex mixtures. Mëdecine often makes use of substances derived from essential oils. [16]

According to the various pharmaceutical bibliographies, essential oils have antiparasitic, antiviral and antibacterial activity, and are also antifungal'[19].

Essential oils and aromatic plants in gënërales are still used for their medicinal properties, in this case their anti-inflammatory and antiseptic properties,

antiviral, antifungal, bactericidal, antitoxic and insect repellent, toning and stimulating, calming, etc.[24]

In order to meet this objective, we propose in our work to approach the extraction of essential oils by the method of hydro-distillation of two spontaneous plants, one from the willaya of Souk-Ahras: ***Origanum vulgare*** and the other from the region of Boukhadra **wilaya of Tebessa**: ***Artemisia herba alba, in*** order to assess their essential oil yields, identify and characterise their chemical components using thin layer chromatography (TLC) and test their efficacy against different species of pathogenic fungi.

Two sections have been devoted to illustrating this:

The 1ere bibliographical section comprises three chapters:

- The first: a collection of bibliographical data on medicinal plants.
- The second describes essential oils and the methods used to extract them.
- The third: provides information on phytotherapy and aromatherapy.
- The2eme practical part comprises two chapters:
- The first: Materials and Methods.
- The second deals with the results and the discussion.

And at the end of our work, we'll draw a conclusion and look to the future.

Partie 01 :

Partie bibliographique

1 part

Bibliography

Chapter 1: Medicinal plants

1-History

The first known text on herbal medicine is engraved on a clay tablet, written by the Sumërians in cuneiform ciracteres 3000 years BC; they used plants such as myrtle, hemp, thyme and willow in filtered decoctions.

The Ebers Papyrus, dating from the 16th century BC, is the earliest known compendium of medicinal plants. At 110 pages, it is by far the most voluminous known from ancient Egypt, and refers to more ancient documents that list dozens of plants along with instructions for their use. [38]

The Greeks and Romans also used many plants. References to them can be found in the works of Dioscorides (a Greek physician from the 1st century), among others.

In Europe, plants were the mainstay of pharmacopoeia until the end of the 19th century and the advent of modern chemistry. Still widely used after the Second World War, they were subsequently supplanted by simpler-to-use synthetic medicines. [18]

In France, the Vichy government abolished the herbalist diploma in September 1941. From 4,500 herbalists in 1941, there are now around ten, while in Germany and Italy there are several thousand.

Since Antiquity, plant specialists have been clearly identified, from doctors to herbalists, and this separation is still in force in other societies around the world: certain plants are sacred, prepared only by the person who fulfils the function of healer. [38]

2-Introduction

Medicinal plants are part of the history of every continent: in China and India, over the centuries, knowledge about plants has been organised, documented and passed down from generation to generation. Today, the use of herbal medicine is enjoying a revival in Western countries, particularly to treat the imbalances caused by modern life, from stress to weight problems.

The use of herbal medicine is becoming part of everyday life, in the form of prevention, and is no longer reserved for the treatment of illness. [4]

In Africa, attitudes vary, partly because of the persistence of witchcraft. Millions of people use traditional medicine first and foremost, and sometimes exclusively, because it remains the most affordable and seems effective. Others prefer Western medicine, because they associate traditional medicine with superstition. One global figure gives an idea of the importance of using traditional medicine: it is estimated that 80% of the world's population uses it for their primary health care. [4]

After fighting traditional medicine for a very long time, doctors and health organisations now recognise the value and effectiveness of treatment with plants. They are also used in the composition of so-called modern medicines.

Artemisia annua, for example, has been used in China for over two thousand years: the active principle of this plant has become the essential ingredient in treatments for malaria. It is now included on the WHO's list of essential medicines, after more than 20 years of research and work. *Artemisia Annua* is now grown in East Africa to supply European markets. [4]

Other African plants are enjoying renewed interest, such as African Geranium, African Plum and *Sutherlandia frutescens*. This plant, which grows only in South Africa, is used to complement therapies to combat AIDS.

On the other hand, other plants are presented as 'miracle' products by certain charlatans who take advantage of the credulity and distress of patients (in particular to treat AIDS).
It is therefore very important for modern medicine and traditional medicine to work together:
- In order to validate and improve traditional remedies.
- So that we can learn from each other and complement each other, taking research to the next level.[25]

3-Definition:

Medicinal or pharmaceutical plants are plants that contain principles that can be used therapeutically and those that, after being dried or processed using different methods, are used in the preparation of medicines. The remedies extracted from these plants are used as raw materials in industry or medicine and are known as galenical preparations (named after Galen, an 11th century physician). [43]

Pharmacognosy is the science that studies the origin, appearance and composition of plants and the way in which they are used. Its aim is also to discover new medicinal plants for use in medicine and cultivation [43]

There are two approaches to phytotherapy, two distinct uses of medicinal plants:

1. The whole plant or part of the plant is used as it is, without having undergone any physico-chemical extraction. The term totum is then used to designate all or part of the plant used. The totum contains numerous families of active ingredients acting in synergy. The patient can take it in the form of gelules containing the plant powder, tablets or herbal tea (extraction of the water-soluble compounds)...
2. Example: harpagophytum root (*Harpagophytum procubens* L.)
3. The whole plant or part of the plant undergoes a physico-chemical extraction. An aqueous or hydroalcoholic extract is obtained, depending on the extraction solvent used. This liquid extract then undergoes a drying stage. This produces a dry extract. The dry extract is concentrated in active ingredients from the same chemical family.

Example: hvdro alcoholic dry extract (*Hypericum perforatum* L.). [47]

4- Source of medicinal plants

Medicinal plants can be found either in the spontaneous wild state, known as "harvested plants" [31], or in the cultivated state, known as "crops".

4-1- Picking plants

Harvesting wild plants covered almost all the therapeutic needs.
Today, it is insufficient for many drugs.
At present, plants are gathered in the wild in the following circumstances:

- If plants are very abundant in the wild.
- For plants that are difficult or impossible to grow.
- In the event of a reduced request. [10]

4-2- Cultivation plants

Growing crops has many advantages:

- From seeds that have been selected from the start, we obtain abundant, good quality raw material.

- Plants all have the same ëtate of maturity when harvested. [33]

5-Habitat of medicinal plants

Medicinal plants can be found at practically every latitude, in the most diverse habitats, wherever the vëgëtal world exists. We find a the spontaneous state, l'Adonis, Thistle benit, Thyme, Rosemary. [12]

- Aconite, Bussero1e and Arnica can be found in mountainous regions.
- In the dirty soils of the seaside, you can find sea wormwood, Ache.
- Marshmallow, mint, etc. grow by the water's edge. [30]
- Anemone, wild thyme, etc. grow on dry hillsides.
- In granitic regions: Foxglove, Lycopodium, Lily of the Valley, Holly, etc.
- In damp pastures, Colchicum and Bistort.
- [12]

In the water, water lilies and watercress.

6-Harvesting plants

Plants are best harvested in dry weather, after the sun has risen and the roses have disappeared.

Plants are also harvested according to their age, the place and the time of harvest. In addition, depending on their different uses, they are harvested differently according to the season.

The medicinal properties and culinary qualities of many herbs and other plants derive from the active and aromatic principles they contain. Harvesting is therefore only appropriate when the organs being sought are richest in these active principles and aromas [9].

7-Drying

The plants are dried in the shade, but in full sun, the action of bright light discolours the medicinal plants collected, and their volatile principles and essential oils are destroyed by the heat. This is especially true for flowers, which are generally the most affected by the action of the sun.

The maximum tempërature allowed for drying aromatic plants or plants containing essential oils is: 30°C.

In all other cases the drying tempërature can vary between: 15°C and 70°C.

It is essential to establish good air circulation to ëvoid fermentation or rotting. [9]

8-Conservation

If the leaves, flowers, seeds, stems, fruits and roots of a plant are to be preserved at the same time, all these parts must be sëchëparëly separated and placed in different jars, usually in earthenware or glass containers, metal boxes may also be used, but care must be taken to ensure that these containers are perfectly clean (sterile) and very well dried; because the worst enemies of medicinal plants are dust.

Each jar should be labelled with the name of the plant it contains to avoid any possibility of confusion; the jars should be kept in a cool, dry place [42].

9- The list of active ingredients in medicinal plants

The active ingredients of a medicinal plant are the components naturally present in the plant, which give it its therapeutic activity. These components are often present in extremely small quantities in the plant: they represent barely a few per cent of its total weight, but they are its essential element.

Active ingredients are found in all parts of the plant, but in unequal amounts. And not all the active ingredients in the same plant have the same properties.

A typical example is the orange tree; its flowers are sedative and its bark is aperitive.
With some plants, only part of the plant can be used. An example is ginseng, where only the root contains tonic substances. [18]

9-1-Alkaloids

These substances are toxic, sometimes in small doses, and have known therapeutic effects. They are nitrogenous organic substances of plant origin, alkaline in character and complex in structure. Alkaloids are found in several plant families and more than a thousand are known. Morphine (1805), strychnine (1818), cafëine, quinine, colchicine, curare, atropine, etc. They pass very easily through percolation. They act directly on the nervous system nervous system with effects on consciousness and motor skills.

Action on the nervous system can go as far as an antispasmodic antispasmodic, and

mydriatic, local anaesthetic or analgesic and narcotic. Alkaloi'des are today namedës after the plant that provided them, always with an "ine" ending.In a дё^гак fagon, alkaloi'des are bitter and usedës as apëritives. [39]

9-2-Heterosides (or glucosides)

These are sugar molecules which are linked either to a **phenol** function or to a nitrate or sulphur derivative which will cause the molecule to produce specific properties.

9-3-Saponins (or saponosides)

Saponosides (soap - soapwort, soap plant, liquorice, white broth, modene) are natural ieterosides whose material is a water-soluble compound that makes it foam like soap water. [25]

9-4-Flavonoids

They are used in the composition of many vëgëtal pigments, in particular yellow and orange pigments (calendula) and also *blue* pigments (cornflower, a great antispasmodic for the face and especially the eyes). Plants containing flavonoids [39] are often linked to the antispasmodic function.

9-5-Anthocyanins (or anthocyanins)

In high doses, anthocyanins are poisons *similar to cyanide*. They are derivatives of cyaniidric acid (the product of combining iydrogen with [39] cyanogen). They are found in blue flowers (cornflowers, violets, mallows).

9-6-Mucilages

They are still hëtërosides. They are large molecules similar to gums which are shaped concretions of sugars. They will spontaneously deposit on the tissues and act [25] as a protector.

9-7-Vitamins :

Amino substances necessary, in small quantities, to maintain life. Vitamins are substances that act in small doses. There are water-soluble and fat-soluble vitamins.
Plants provide almost all vitamins. Some plants are rich in them
(Ex: Lemon--> vitamin C; Watercress--> vitamins B1, B2, C, E). Examples: Rosaceae7, Rutaceae 8, Fabaceae 9. [39]

9-8-Tannins :

Tannin is a **phenol** associated with a sugar. One of the basic tannins is gallic acid. They precipitate (agglutinate, coagulate) proteins and gelatin, which is much rarer.
In addition, they can be used in cases of alkaloid poisoning, as it precipitates them and renders them harmless (except for morphine, cocaine and nicotine, no interaction). But if the dose is

forced, the excës of tannin HЬёге again the toxic substance and cause a secondiëme inflammation. [39]

10-Preparation methods in phytotherapy

Depending on the therapeutic effect sought, traditional use and then research have developed processes for treating plants so that only the molecules of interest are retained, for use locally, by mouth or by injection.

In preparations, the composition of a **remedy** may combine different plants. Herbal teas, poultices applied directly to the skin, syrups, alcoholic or aqueous solutions, essences and oils are the most common forms of remedy. Here, we will look at the main galenic forms available and their manufacturing origins[39] **10-1 - Herbal teas**

Herbal teas are obtained by maceration, digestion, infusion or decoction using water.

-*S* **Infusion:** This involves pouring boiling water over the plant, covering it and leaving it to cool for 2 to 15 minutes. It is suitable for fragile plants (flowers and leaves).

Decoction: This consists of boiling the drug in water for 15 to 30 minutes. It is suitable for "hard" plants (bark, roots, fruit and certain leaves).

Maceration: This involves keeping the plant in contact with water (at room temperature) for 30 minutes to 4 hours.

Digestion: The plant is kept in contact with water (at a temperature below boiling, but above room temperature) for 1 to 5 hours.

10-2-Powders

Prepared by pulverisation followed by sieving, they are used directly in the composition of geiules, but are also used in the manufacture of other gaianic forms such as extracts and tinctures [39].

10-3-Extracts

Extracts are obtained by treating the plant in a vaporisable solution (ether, water, alcohol, etc.) using various extraction processes (maceration, digestion, infusion, digestion, leaching) and then evaporating these solutions to obtain a fluid, soft or dry consistency. They are therefore classified according to their consistencies. [28]

10- 4-Tinctures

They are obtained from dried plant powders and their alcoholic strength varies according to the type of drug. It may be 60° (highly soluble active ingredients), 70° or 90° to 80° (e.g. resinous products and volatile oils). [28]

10-5-Spirits

These are tinctures prepared from fresh plants that have not undergone the effects of desiccation.

10-6- Alcoholates

They are obtained by distilling the volatile principles of plant substances in contact with alcohol. They are always colourless and indestructible, but must be kept in [39] tightly stoppered bottles.

10-7- Intraits

They are only justified in cases where the active ingredients of a drug (e.g. horse chestnut, valerian, sage, colchicum, etc.) are likely to be degraded after harvesting, necessitating a "stabilisation" operation (elimination of inactive and/or degrading products such as ferments) using water.

10-8- Essential oils (EO)

They come in two forms:

A) Solid EOs, also known as "petrol camphors".

B) Natural or dissolved liquid EO (e.g. rose EO).

C) EO officinalis is obtained by steam distillation, expression or incision.

They are classified according to their colour (blue, yellow, green-brown or colourless) or their chemical composition [39] (hydrocarbon, sulphur and oxygenated hydrocarbons).

10- 9- Distilled waters or hydrolats

Hydrolats are obtained by distillation (with water) of plant powder or parts of plants (flowers, flowering tops). Distilled waters, or hydrolats, are very [39] fragrant because the EOs are suspended in water (in the case of solids).

11-Purpose of use

In traditional phytotherapy, plants can be used fresh, which is not always possible, or dried, and then used in various preparations to preserve their active ingredients. They are administered in the form of alcoholic tinctures, maceras, herbal teas, compresses, balms, etc. (Note: only the water-soluble active ingredients of the plant can be extracted from herbal teas).

More recent processes can be used to obtain all the active ingredients, the most rudimentary of which involves fine grinding (cryo-grinding, for example) of the plant after drying to obtain a total plant powder. This powder can then be presented in the form of tablets, capsules, sachets, etc.

Another process, extraction, optimises the herbal tea principle using cutting-edge technology to obtain a purified powder concentrated in active ingredients. After the plants have been coarsely ground, the powder is immersed in a natural solution, the solvent (water or alcohol), adapted to the physicochemical properties of the active ingredients being sought. This crucial phase enables all the active ingredients to be isolated and their synergistic action to be preserved. The liquid thus obtained then undergoes a filtration phase to eliminate non-assimilable solid compounds such as cellulose. This is followed by a moderate vacuum drying phase, which gradually eliminates the solvent and any contaminants. The dry extract thus obtained has the following advantages:

- Concentration of active ingredients
- Uniformity of active ingredient content (independent of natural variations, location, season, etc.)
- Maximum bioavailability by eliminating non-assimilable compounds
- Health safety: absence of microbiological and chemical contamination

In Europe, certain plants are recognised as medicinal plants and, like medicines, are subject to marketing authorisation. As a result, products extracted from certain plants are only sold in pharmacies, and sometimes only on prescription. [28] **11-1/ Internal use**

Fumigation

This involves using vapours charged with the plant's active ingredients, for example boiling eucalyptus leaves in a room to be disinfected.

There are also wet fumigations, in which a plant is boiled: either an inhaler is used, or the technique of covering the head with a terry towel, with the face placed over the bowl of steaming water containing the plants. [39] **11-2/ External use :**

11-2.1. Skin :

A. **Compress**

This involves applying gauze soaked in decocte, infuse or macërë to the areas to be treated.

B. **Cataplasm**

This is the preparation of the plant, which is pasty enough to be applied to the skin for therapeutic purposes.

The plant can be ground, chopped hot or cold, or mixed with flax meal to obtain the right consistency.

The poultice soothes muscular pains and neuralgia, relieves sprains and fractures and helps to extract pus from infected wounds. [39]

C- Lotions :

Lotions are water-based preparations of plants in infusions, decoctions or diluted tinctures, which are dabbed onto the epidermis in irritated or inflamed areas.

D- the Baths

In the bath, simply pour an infusion or decoction of plants into the water.

This can be a full bath or a partial bath. It is prepared by adding an infusion, decoction or macerate to the bath water.

1-Complete bath: It can be invigorating or, on the contrary, calming ...

2-Partial bath: a distinction is made between :

- The sitz bath, or ano-bladder bath, is indicated for the treatment of haemorrhoids and anal fissures. The cold sitz bath has a decongestant effect on the pelvis.
- Footbaths and handbaths are indicated in the event of excessive sweating of the feet or hands. [39]

11-2.2 Mucous membranes

A- Gargle

The mëdication, consisting of an infusion or decoction as hot as possible, is used to rinse the mouth, throat, pharynx, tonsils and mucous membranes. The gargle should never be swallowed.

B-Mouthwash

Infused, decocted or macerated, it is used to treat mouth ailments such as mouth ulcers. [39]

C- Eye bath

It is carried out using a beaker filled with an infusion or a decanter; it is [39] essential to filter the solution before use.

12- Precautions for use of phytotherapy

Some plants contain active ingredients that can be extremely powerful, while others are toxic in small doses. The fact that only plants are used does not mean that they are harmless; the free cultivation of certain plants is prohibited in certain countries, the most common example being poppies, the cultivation of which is regulated in France and is intended solely for the pharmaceutical industry.

Pharmacology recognises the beneficial action of certain plants and therefore focuses on extracting the active ingredient from these plants. Consumption of the 'raw' plant leads to the consumption of products other than the active ingredient, making it impossible to know the exact dose of active ingredient ingested, leading to the risk of under- or overdosing. For some phytotherapists, the other ingredients will attenuate the side effects by interacting with each other. For example, the distillation of lavender yields more than 200 different molecules, including cetones and coumarins, which are less toxic than if used alone.

The composition of a plant may vary from one spëcimen to another, depending on the soil, growing conditions, humidity, temperature, sunlight, which will determine what is known in aromatherapy as the chemotype. [38]
Similarly, plants of dubious origin should not be used, as pollution factors, harvesting and methods of preservation and storage can alter the properties of the plants.

13- - Interaction

Taking herbal remedies and medicines at the same time can lead to interaction between the two remedies and the appearance of sometimes serious side effects.
For example, St John's Wort can inhibit the effect of drugs such as dioxin, theophylline, anti-vitamin K anticoagulants, oral contraceptives and certain antidepressants, or less commonly used drugs such as cyclosporine, treatments for HIV infection (AIDS) such as amprenavir or indinavir, or certain anti-cancer drugs.
Lastly, we should note that some people present phytotherapy as a "natural" method, an argument that is often based on advertising or a fashion effect that plays on an ambiguity: natural equals "beneficial" and "harmless" (whereas nature is neither good nor bad, and death, disease, venoms and toxins are all natural). It is estimated that 5% of intoxications are caused by plants, sometimes by phytotherapeutic preparations. [18] such as wolfsbane

14- Indications

However, new, rigorous research protocols are now being developed which respect the specific characteristics of plants (synergy, consideration of trace elements, vibratory action, etc.). For example, consideration is being given to studying the physiological responses to plant treatments (stimulation of blood circulation, expectoration, diuretic effects, influence on digestion, etc.) rather than statistically evaluating their effects on morbidity. [25]

- Beta-carotenes for tanning
- Wheat and soya to combat ageing
- Heavy legs
- Constipation
- Diarrhoea
- The cëphalëes
- Urinary problems
- Bronchial syndromes
- Back pain
- Arthritis
- Rheumatism
- Rhinitis
- Weight loss
- Stress
- The dëpression
- Migraines
- Insomnia
- Arterial hypertension
- Lymphatic drainage

15 - Contraindications

Not everything that is 'natural' is harmless. Some plants are simply poisonous and others can be harmful in interaction with other plants, medicines or supplements. Most medicinal plant monographs indicate potential harmful interactions for each. [28]

16-Examples of medicinal plants

16-1- 'Armoise

> Systematic position of the species *Artimisia herba alba :*

Kingdom: Plantae

Class: Equisetopsida

Family: Asteraceae

Order : Asterales

Genre: Artimisia

Species: *Artimisia herba alba*

Figure 01: Mugwort [50]

> History

Many books on Greek and Roman mythology state that mugwort, the "mother of herbs", was held in high esteem as a medicinal plant. In the Roman Empire, mugwort was supposedly planted along the roadsides so that legionnaires would not have tired legs when travelling. This belief persisted for centuries, well into the Middle Ages. The scholar William Coles wrote in 1656: *"...by putting mugwort in one's shoes in the morning, one can travel leagues and leagues until noon without fatigue".*

Magic powers were frequently attributed to mugwort. A bunch of mugwort hung under the eaves was said to protect the house and all its inhabitants from epidemics and lightning. Historically, there is no doubt that the botanical genus Artemisia - Douglas mugwort *(Artemisia douglasiana), for example,* is used as a medicinal plant in traditional Chinese medicine - has lost none of its legendary magical virtues to this day. [11]

> Description

Common names: *Artemise - Absinthe sauvage - Herbe de la Saint-Jean - Couronne de Jean-Baptiste - Herbe de feu - Herbe aux cent gouts - Tabac de Saint-Pierre - Sourcil de loup.*

Mugwort is a perennial herbaceous plant 0.50 to 1.50 m high, with a thick, woody stump and a rigid, upright, reddish, branching stem.

The sessile, dark green leaves, hairless above and whitish below, are cut into acute lobes and give off an aromatic scent reminiscent of wormwood when rubbed between the fingers. The flowers, which are greenish yellow, tubular and in small ovoid heads, are grouped in large panicles of loose epis and the fruits are oblong, hairless akenes.

It grows on wasteland, rubble, embankments, stony ground, along railway lines, forest paths and river banks, up to an altitude of 1,600 m. Common in all temperate regions of Europe, northern Asia, Africa and America. [47]

> **Botany and culture**

Common mugwort, a "weed" that grows in virtually any soil, is an imposing perennial plant that can reach heights of over 1 metre. It can be propagated by sowing or by cutting young shoots. This aromatic plant has woody stems at the bottom and medium-green, silverëesque, downy leaves on the underside.

The plant tolerates winter frost very well. We recommend pruning it to 15 cm from the ground in early autumn so that new shoots can form before winter. Its purple flowers add a discreet splash of colour to any garden in the summer months.

The leaves are used first and foremost. But in some regions, notably China, the young flowering shoots are also used to flavour delicate dishes. This solitary plant of imposing size prefers sunny locations, preferably not too damp. [11]

> **Use in natural medicine**

We recommend using this bitter plant sparingly, both in medicine and in cooking. After all, *"the best is often the enemy of the good"*. But mugwort tea frequently relieves stomach and intestinal complaints. The beneficial effects of this plant on digestion, as well as its soothing action on menstrual disorders and skin problems, have always been highly appreciated. It should be noted, however, that common mugwort should not be used during pregnancy or while breast-feeding. [34]

16-2- Thyme

> **Systematic position of the *Thymus vulgarus* species**

Kingdom: Plantae

Family: labia

Order: Lamiales

Genus : *Thymus*

Species : *Thymus vulgarus*

Figure 02: Thyme [49]

> **History**

Thyme was already an important plant in ancient Egypt. The scholars of this highly developed culture used it to embalm the dead. It is undisputed that thyme has preservative properties. It protects paper from mould and is still used today in embalming fluids ийНзёз to preserve botanical and anatomical collections.

The valiant warriors of ancient Rome took thyme baths to fortify themselves and keep their courage up during battle. *Thymus* comes from the Greek word *thymon*, meaning courage, strength. This is also why, in the Middle Ages, damsels embroidered sprigs of thyme on the

scarves they gave their valiant knights when they set off to distant lands.
The hëros ëcossais tëmëraires also knew how to profile the virtues of thyme and drank a tea of this aromatic herb to give them courage and strength and ward off nightmares.
The Greeks burnt thyme as incense in temples and sacred places. In more modern times, thyme was considered the "enemy of poison" and was used to fumigate patients' rooms. This plant, which astrology attributes to Venus, is used as incense, among other things to purify places where rites are held and to promote clairvoyance.
This familiar herb, used in cooking, medicine and as incense, originated in the eastern Mediterranean, where it grows in dry, rocky soil. Benedictine monks brought it north in 800 AD and began growing it in their monastery gardens. This valuable herb with its many uses very quickly found its way into gardens and gradually spread throughout Europe. [11]

> **Description**

Thymes are small, bushy plants, either spreading (common thyme) or more sparse (wild thyme), typical of the Mediterranean region where they grow naturally.
Thyme is grown all over the world, some varieties tolerating the cold and damp, others the tropical heat. However, thyme from the garrigue has the highest concentration of essential oil, which no doubt helps the plant to survive the aridity and summer heat of the Mediterranean environment. Botanists distinguish between numerous species of thyme, which are quite difficult to differentiate and are often typical of a particular region or country (endemic species). [48]

> **Botany and culture :**

Common thyme, frigoule, farigoule, barigoule, pote

Thyme vëritaBle is a small sub-shrub that grows to a height of up to 40 cm. It is a member of the Labiacëes family. Its square, erect stems are short and hairy, and whitish in colour.
Its small, dark green leaves, 4-10 mm long, are elliptical to oblong and short-stemmed, cottony and greyish beneath and smooth above. They are tightly rolled at the edges. The entire plant has an aromatic scent and flowers from June to September. Its pretty flowers are purplish to dark pink in colour and are much appreciated by bees.
There are different types and variëtës of thyme, all of which have the same habitat and care requirements. Thyme needs a sunny spot with light, draining soil if it is to fully develop all its virtues. In spring, it can be given a little compost or complete organic fertiliser, and again in autumn, so that the shrub can gather all its strength for the winter. True garden thyme is relatively resistant to the cold in our regions, but it is advisable to protect it in winter in exposed spots by covering it with fir boughs, for example. It does not need pruning if it is harvested regularly; if not, it is advisable to cut back the shrub by about a third in autumn, taking care not to cut too much into the old wood. This versatile herb can be propagated by cutting, sowing or dividing the clump.
This plant is a must for any herb or scented garden, and also has its place in rock gardens. It loves the proximity of stones, which release stored heat, and is very decorative in a garden or pot with its pink flowers. Aromatic thyme also looks great in a small decorative herb basket. It goes well with other Mediterranean herbs such as sage, rosemary and lavender. [11]

> **Use in medicine**

Thyme played an important role in European folk medicine, helping to combat respiratory ailments and strengthen the body. Hildegarde Von Bingen also appreciated the strength and antibacterial and antiviral properties of thyme, which she used extensively. Thyme was also said to have beneficial properties against stomach and intestinal disorders, gout and rheumatism, as well as coughs, sore throats and infections.

In cosmetics, thyme is used fresh or dried for facial steam baths and skin cleansing. [36]

Chapter 2: Essential oils

1-Definition:

Essential oils, also known as essences, are mëlanges of aromatic substances produced by many plants and present in the form of tiny droplets in leaves, fruit skins, resin, branches and wood.

They are present in small quantities in relation to the vëgëtal mass: they are odorous and highly volatile, i.e. they evaporate rapidly into the air.

It is important to distinguish between essential oils, fixed oils (olive oil, etc.) and fats contained in vegetable oils. Indeed:

J Only essential oils are volatile, which distinguishes them from fixed oils and fatty oils.

They are frequently combined with other substances such as gums and resins.

Moreover, they themselves tend to resinify through exposure to the air. [1]

2-Physico-chemical properties of essential oils

2-1-Physical properties :

According to **Abdelouahid D.**, oils have a number of physical properties in common:

- They are soluble in alcohol, cther, chloroform, fixed oils, emulsifiers and most organic solvents, and sparingly soluble in water, to which, however, they impart their odour.
- Their boiling point varies from 160°C to 240°C.
- Their density is generally lower than that of water, varying from 0.75 to 0.99 (sassafras, clove or cinnamon essential oils are exceptions).
- They have a high fraction index
- They are dextrorotatory or, rarely inactive on polarised light.
- They dissolve fats, iodine, sulphur and phosphorus and reduce certain salts.
- They are perfumes and have a limited shelf life.
- Are very weatherable and sensitive to oxidation (but do not go rancid)
- These substances have an oily consistency, are more or less fluid, even retinoid, and are highly odorous and volatile.
- *At* room temperature, they are generally liquid, colourless or pale yellow, although there are a few exceptions, such as azulene essential oil, which is blue in colour.
- These are stimulant products, used both inside and outside the body, sometimes pure, usually dissolved in alcohol or a suitable solvent. [15]

2-2-Chemical properties :

The chemical composition of essential oils is highly complex, and they can contain up to several hundred different chemical molecules.

The constituents of essential oils belong almost exclusively to two groups characterised by their distinct biogenic origins:

J The terpenoi'des group.

S The group of aromatic compounds derived from phenylpropane. [1]

2-2-1-Terpenes group

In the case of essential oils, only the most volatile terpenes will be encountered:

A- Mono terpenes :

The simplest components of the series, monoterpenes are produced by coupling two "isoprenic" units.

They can be acyclic (myrcene, ocimene), monocyclic (a- and y-terpinene, p- cymene) or bicyclic (pinene, camphene, sabinene). They sometimes make up more than 90% of essential oils.

Structural variations justify the existence of numerous molecules:

- Alcohols (geraniol, a-terpineol, borneol, trans-trans-farnesol)
- Phenols (thymol)
- Aldehydes (citronellal)
- Cetones (carvone, в-vetivone)
- Esters (Cedryl acetate)
- Ethers (1,8 cineole). [15]

B- Sesquiterpenes

A large number of sesquiterpenes are common constituents of the essential oils of higher plants, and may play a part in the pharmacological properties attributed to these volatile fractions.

Biologically, a good number of sesquiterpene structures are phytoalexins, others seem to act as growth regulators, and others attract insects or act as anti-nutritive factors when they encounter them. [15]

2-2-2-Aromatic compounds

Phëhylpropane (C6-C3) dërivës are much less common than prëcëdents. They are very often allyl- and propënylphënols, sometimes aldehydes. (C6-C1) compounds such as vanillin or methyl anthranilate can also be found in essential oils. [15]

2-2-3-Composites of various origins

Essential oils may contain various compounds resulting from the transformation of non-volatile molecules and which can be entrained by water vapour.

Low molecular weight organic acids Acetic and Valenic isovalenic acids.

- Low molecular weight cetones.
- Volatile coumarins e.g. bergaptenes.
- Nitrogen or sulphur compounds.
- Compounds derived from the degradation of fatty acids, for example the formation of jasmonic acid and its esters.
- Compounds derived from the degradation of terpenes, such as ionones, result from the auto oxidation of carotenes. [36]

Essential oils have a number of chemical properties in common:

- They oxidise in the light and resinify as they absorb oxygen. At the same time, their odour changes, boiling increases and their solubility decreases.
- Absorb chlorine, bromine and iodine with heat release.
- Change to a yellowish colour through oxidation, linked to the formation of resin. [ ][23]

3-Biosynthesis:

Given that terpenes are important characteristic constituents of most essential oils, for this reason we will limit our study to the biosynthesis of this chemical group. Terpenes are a family of compounds widely distributed in the plant kingdom. In the case of essential oils, only the most volatile terpenes will be encountered, i.e. those whose molecular weight is not too high, such as monoterpenes and sesquiterpenes, which can exist either in the form of hydrocarbons with the crude formula $(C_5H_8)_n$ or in the form of oxygenated derivatives of these hydrocarbons called : terpenoids include alcohols, aldehydes, esters, ethers, cetones, phenols and oxides. [37]

An important feature of the chemical structure of terpenes is the presence in their skeleton of the recognised 5-carbon isoprene base unit (C_5H_8).
This isoprene is the basis of the concept of the "isoprenic rule", which considers isopentenyl diphosphate (IPP), designated in the form of active isoprene, to be the true precursor of the terpene molecule, hence the name isopprenoid under which it can also be designated. [37]
The first stage in the biosynthesis of the IPP molecule involves the condensation of two acetyl CoA molecules to form acetoacetylCoA. This reaction is first catalysed by an acetyl CoA thiolase coupled to an Fe(2) quinone, HMG CoA synthetase then catalyses the binding of a third acetyl CoA to give B-hydroxy-B- methylglutaconyl coenzyme A (HMG CoA). The reduction of the acid function (involved in a thioester bond) to an alcohol is then catalysed by HMG CoA reductase to give mevalonic acid. [37]
A diphosphate group will then be attached to the primary alcohol function of the mevalonic acid, the mevalonic 5-diphosphate formed will react with a third molecule of ATP, this reaction provides an unstable compound which spontaneously decomposes by losing a tertiary alcohol function and the free carbonyl group leading to the formation of IPP which is the biologically active form of the isoprenic molecule.
The biosynthesis of terpenes involves the addition of the isoprene unit (IPP) with isomer to form geranyldiphosphate (GPP: C10) which is the precursor of all mono terpenes, GPP will condense with another IPP unit to form the di phosphate of [37].
farnesyl (FPP: C15), the source of sesquiterpenes.

4-Distribution, location, formation

Essential oils are found in various botanical families. They are found in all living parts of the plant and are formed in the cytoplasm of spëcialisëes cells. [20] **4-1-Distribution**
Essential oils are widely distributed throughout the vëgëtal world. They are found in appreciable quantities in around 2,000 species divided into 60 families. By way of example, we will mention some of them:

- Asteraceae (mugwort, chamomile, dandelion)
- Lamiaceae (basil, lavender, mint, patchouli, romani, thyme)
- Lauraceae (camphor, cinnamon, laurel)
- Myrtaceae (eucalyptus, cloves)
- Apiaceae ^дё^^, carrots, chervil, parsley...)
- Abietaceae (ёркёя, pine, fir.)[20]

4-2-Location

These essences can be found in all living parts of the plant, such as :

- The flowering tops of basil, lavender or mint.
- Ginger or cinchona rhizomes.
- Vetiver roots.
- Nutmeg seeds.
- The fruit of aniseed, fennel or orange.
- The flowers of the rose.
- Eucalyptus, laurel or patchouli leaves.
- The bark of the cinnamon tree.
- Camphor or sandalwood.

They are often more concentrated in twigs, flowers and seeds. In the same plant, these oils may exist in different organs at the same time. The chemical composition may vary from one organ to another. In the case of the lemon tree, for example, the flower and fruit provide essences with different chemical compositions. [20]

4-3-Training

Essential oils, produced by higher plants, are elaborated by large secretrices found on almost all parts of the plant (stems, leaves, flowers, etc.). They are sëcrëtëes within the cytoplasm of certain cells where they resemble small droplets like most lipophilic substances. [20]

They are then stored in cavities resulting from the fusion of several cells.

The synthesis and accumulation of essential oils are accompanied by special histological structures that differ and vary according to botanical family. Examples include :

- The external secretory hairs of the Lamiaceae are located on the lower epidermis of the leaves.
- The internal secretory hairs of Myrtaceae are found in various plant organs and rhizomes.
- Secretory pockets of Myrtaceae or Rutaceae.
- Isolated secretory cells of the Lauraceae.
- The secretory channels of Apiaceae and Conifëres.
- Secretory cells modified into oily sacs or tubes.

These reservoirs of oil are formed either by schizogony (division and separation) or by lysogeny (fragmentation and disintegration).

The secretory cells are often located on or near the plant, which facilitates their emission.

When the temperature is high enough, the essences pass through the cell wall and cuticle in the form of vapour towards the outside, releasing the fragrance of the flowers.· [16]

5-Role of essential oils in plants

Essential oils emitted by plants in vapour form have multiple functions in nature.

At present, it is often difficult to specify them in all cases. Nevertheless, it seems likely that they have an ecological role.

Experimentally, it has been established that they are involved in plant-animal interactions. They are therefore a means of communication (chemical language).

Some species attract insects and encourage pollination. Others defend plants against predators (herbivores, insects, micro-organisms, etc.).

They can paralyse the masticatory muscles of attackers due to the toxic and inappropriate properties of the substances they contain.

They protect crops by inhibiting the multiplication of bacteria and fungi, as well as inhibiting germination and growth.

In addition, they exude a variëtë of taste and smell into the atmosphëre. This is why many of them are used as flavourings and condiments in cooking.

Finally, it should be noted that for plants in desert regions, the oil vapours saturate the air around the plant, maintaining a certain humidity that prevents the temperature from rising excessively during the day and falling during the night.

However, the function of essential oils within plants is still a rather obscure phenomenon. The attraction, repulsion and release effects observed are due to the complexity and structural variety of the substances present in the essence. [32]

6-Chimiotaxonomy

For some time now, the convergence of chemistry and botany has given rise to a new discipline, 'chemotaxonomy', which is developing spectacularly.

He also believed that there was a definite relationship between the systematic position of the

plant and the chemical substances it contained. [15]
Chemotaxonomy can therefore be defined as a tool available to the botanist to differentiate between species, subspecies and hydrides with the same morphology; these are chemotypes and chemotaxonomy is based on an analysis giving the quantities of the different constituents of an essential oil during its biological cycle.
In a way, the essence represents the 'personality' of a plant which, as a living being, is unique and cannot be reproduced. It carries within it the imprint of the plant from which it comes and, if it is pure and extracted in a correct and respectful way, it is extremely powerful. [15]
Chemo type is taken to mean the chemical type of a plant classified according to the majority constituent of its essential oils. Mixed or intermediate chemo-type refers to the chemical type of a plant classified according to two or more majority constituents of similar preponderance. Two researchers worked on two samples of calament, and found that one of the two essences was of one chemotype (Pulegone - chin - isomenthone), while the other sample had a different chemotype (earvone - 1.8 cineole_limonene). Since each chemo type has its own chemical profile, the resulting therapeutic applications are obviously different. [15]

7- Extraction methods for essential oils

Essences are the most precious things plants produce. Since time immemorial, people have been trying to find ways of extracting essences from plants so that they can be used to make medicines, cosmetics and perfumes.
Thus, after harvesting, and depending on the part of the plant to be extracted (whole plant, flower petals, flowers, leaves, roots or fruit), the extraction process used is different and consequently the composition of the extract is different.
What introduces this diversity is, firstly, the variety of raw materials and, secondly, the considerable sensitivity of certain perfumes, which only require the use of mild means without the intervention of overly energetic chemical agents. [1]
The choice of the type of extraction must make it possible to :

- Remove the aromatic species with the highest yield from the plants.
- Preserve the most delicate fragrances as intact as possible.

The method used to obtain essential oils therefore plays a decisive role in oil yield and composition.
Among the various extraction processes, we would highlight the following:

7-1-The pressure technique

Perhaps this is the oldest method, as the Egyptians used pressure with a sack to extract the essence from flower petals. This method consisted of crushing the fragrant parts of a freshly cut plant, then enclosing them in a linen bag that was twisted using two sticks threaded through two rings at the end of the bag. The essence filtered through the cloth and was collected in a container underneath. [7]
The essential oils of hesperides or citrus fruits such as lemons, bergamots, oranges, etc...., are extremely important in the perfume and cosmetics industry, as well as in aromatherapy. The essential oil is contained in the oleifëres sacs of the fruit peel (pericarp or flavedo of the fruit), still referred to as zest. These essences are very fragile products due to their terpenic and aldëhydic nature. For this reason, all the processes used to obtain them are cold processes which are generally referred to as "cold expression processes".
To do this, machines are used to extract the essential oil by creating zones of sufficient compression and depression in the bark for the plant oil to be released. [1]

7-2-Extraction using solvents

7-2-1-Extraction using volatile solvents

As for extraction using volatile solvents, this has the disadvantage of making it more difficult to recover essential oils, but on the other hand it has the advantage of being able to process plants with an infinitesimal percentage of odorous principles.

This method of production, which leads to concretes and absolutes, should be avoided for essential oils intended for therapeutic use. This technique is currently used all over the world, either to obtain products that cannot be extracted by any other process, or to obtain much higher yields.

The most commonly used solvents, subject to specific restrictive legislation, are aliphatic hydrocarbons: petroleum ether, hexane, but also propane and liquid butane (under pressure). Although benzene is a good solvent, its toxicity is increasingly limiting its use. We also use halogenated solvents (chlorine and fluorescent derivatives of methane and ethane) and ethanol, the latter being used mainly to obtain absolutes and lava resinoids. After extraction? The solvent is distilled. At the end of the process, the solvent that has soaked the plant mass is recovered by injecting steam into it. The choice of solvent is influenced by technical and economic parameters: selectivity (solvent power with regard to the odorous constituents), stability, chemical inertness, boiling temperature not too high to allow total elimination, not too low to avoid losses and therefore higher costs; safety of handling. [1]

The first phase of this process involves digesting the flowers in a container called a "digester". The solution is then distilled, to recover both the solvent and the essence, which at this stage is known as the concrete essence.

This product contains traces of foreign matter and should be treated with pure alcohol, although this will not dissolve all the impurities. To eliminate them, the alcohol must be heated to a high temperature to concentrate the residues. This is when the solution is called extract. After a further distillation, it finally becomes an essence known as absolute.

The major disadvantage of solvent extraction is their lack of sëlectivity, which means that many lipophilic substances can be found in the concretes (fixed oils, phospholipids, carotënoids, certain coumarins) and require further purification, necessitating laborious consëcutifs procedures in order to obtain a product of absolute purity. Another drawback lies in the toxicity of solvents: restrictive rules of use, problems of residues in the final product. [7]

7-2-2-Depletion using fixed solvents

A- Effleurage or cold fat extraction

It consists of making the odorous principles soluble in fatty substances.

This method was widely used in ancient Persia to produce rose ointment, as well as in Egypt. A layer of animal fat was spread on grids and sprinkled with petals of the most delicate flowers. As the flowers wilted, they impregnated the fat with their perfume, and they were constantly replenished with fresh flowers until the fat was saturated with perfume. The Egyptians used to fill a cone with perfumed fat and fold it over their heads: with the heat of the body, the fat gradually melted, releasing the fragrance of the essence. [7]

At present, this technique is rarely used due to its high cost, and is reserved for certain extremely delicate flowers, such as jasmine, tuberose and orange blossom. The substance thus obtained has a very high concentration and is then diluted and treated with other solvents that dissolve the fatty matter. [1]

8- Extraction by maceration in hot fat

The technique known as "digestion" is carried out while hot, by immersing the plant organs in

the melted fat. The resulting product is a floral ointment.
Washing the ointment with strong alcohol produces an alcoholic extract. The alcohol is removed, as in the case of enfleurage, by concentration under vacuum at low temperature. This produces a "maceration absolute". [1]

7-3-Distillation

Distillation is suitable for oils with a high volatile component and is based on the characteristic that these components can be easily transported by moving particles of water vapour. Distillation is still the most widely used method for obtaining aroma compounds because it produces volatile substances that can be easily analysed by gas chromatography and requires relatively simple technology, which means lower costs and easily controllable reproducibility.
Most essential oils are obtained by distillation, with the exception of hesperide essential oils (lemon, orange, etc.) and cade oil. [16]
The steam penetrates the plant tissues and vaporises all the volatile substances; a sufficient quantity of steam allows the plant essences to be isolated. There are three main methods of distillation:

7- 3-1-Hydrodistillation

This method involves directly immersing the plant material to be treated (intact or possibly ground [turbo-distillation]) in a still filled with distilled water, which is then brought to the boil. The heterogeneous vapours are condensed on a cold surface and the essential oil separates by density difference. [1]

Figure03: Hydro distillation set-up (Clevenger). [13]

7-3-2-Steam distillation (or Vapo-hydro distillation):

The plant material in this case is supported by a grid or perforated plate placed at a suitable distance from the bottom of the still. The lower part of the still is filled with water. The level of the water must be such as to avoid any contact between the water and the plant. [16]
The particles of water vapour, moving upwards, burst the cells containing the petrol and take the odorous molecules with them. The vapour then passes through a refrigerated container where the temperature drops, causing the oily molecules to detach from the vapour parti cules, which condense into water. Oil and water separate because of their different specific weights: oil will float on water because it is lighter.
The distillate, which contains the water-soluble components of the essence (aromatic water), is poured through a tap to obtain the pure essential oil.

Sometimes the oils obtained undergo further distillation, called rectification, to eliminate certain particularly irritating substances, as is the case with thyme, or they are redistilled at different temperatures to obtain constituents such as white camphor.
It should be noted that it is possible to work at moderate overpressure in order to shorten the treatment time, limit alteration of the essential oil constituents and save energy.
Distillation has a significant effect on the characteristics of essential oils, and hydrolysis reactions are among the most important. This is why oils extracted in small quantities in the laboratory are qualitatively and quantitatively different from those obtained industrially. [1]

7- 3-3-Direct steam distillation: "separate generator" or "vapo distillation

This method is similar to that described above, except that this time there is no water at the bottom of the still. Saturated or superheated steam, at a pressure generally higher than atmospheric pressure, is introduced into the bottom of the still by a pipe system and passes through the plant mass from bottom to top. The steam comes from an independent boiler.
The essential oils obtained by distillation never exactly represent the aroma and fragrance naturally present in the plant.
Extraction, the second technique for obtaining aromatic extracts, can solve some of the problems associated with distillation. [16]

7-4- Extraction with supercritical co2

The originality of this technique lies in the solvent used. It is supercritical CO2. In the supercritical state, CO_2 is neither liquid nor gaseous, which gives it good extraction power that can be adjusted at will by adjusting the temperature at which it is used. Supercritical fluids such as CO_2 are good solvents in the supercritical state and poor solvents in the gaseous state. The advantages of this process are as follows:

- co2 is chemically inert, natural, non-toxic and inexpensive, and is used at low temperatures.
- The separation between the solvent and the extract is easy (simple dëtente which ratëne the co2 a
ГёШ gaseous), almost completely and inexpensively.
- Operating costs, on a pilot laboratory scale, are reduced (the co2 is continuously recycled).

Extraction with supercritical co2 is an interesting technique that brings new olfactory notes, but it is still expensive to install and the equipment is still invasive. [21]

8-The different techniques for analysing essential oils

Various techniques can be used to identify and characterise the compounds in essential oils:

8-1- Chromatographic techniques

- Separation and quantification techniques
- Separation between two phases: Interaction between the solute and a stationary medium.

8-1- 1- Thin layer chromatography (TLC)

Quite an old method (over 50 years old), not very efficient, but it continues to be used a lot because of :

- A rather unsophisticated piece of equipment
- Very low cost of ownership

Rapid processing of a large number of samples (for routine analyses). [21]

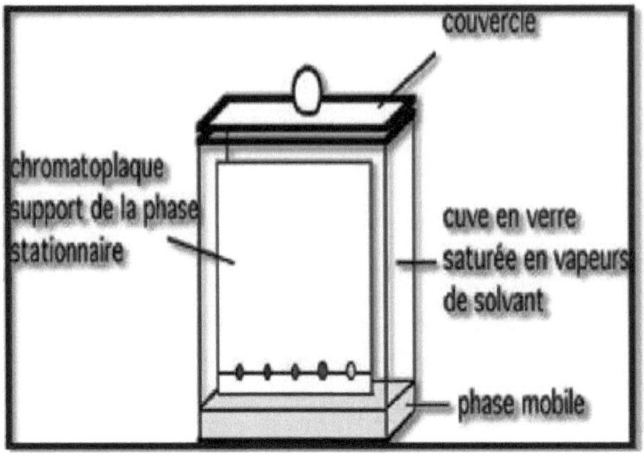

Figure 04: Thin layer chromatography (TLC) [21].

8-1-2 - Gas chromatography (GC)

- Sëparation of gaseous mixtures as a result of equilibria between a mobile gaseous phase and a stationary phase which may be liquid (partition chromatography - GLC) or solid (adsorption chromatography - GSC).

Naturally volatile molecules, but also molecules which, through dërivatisation reactions, are rendered volatile at temperatures which do not cause them to decompose.· [7]

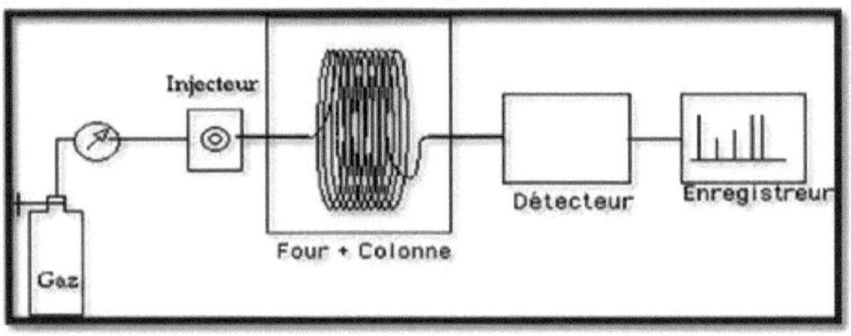

Figure 05: Gas chromatography (GC) [21].

8-1- 3- Liquid chromatography (LC - HPLC)

The areas of application are very varied:

- Separation of mineral ions,
- Small organic molecules
- Macromolecules (proteins, poly nucleotides, organic polymëres).
- For each category of substance there are one or more appropriate chromatographic modes. It's up to you to decide which one is best suited to the job you want to do. [7]

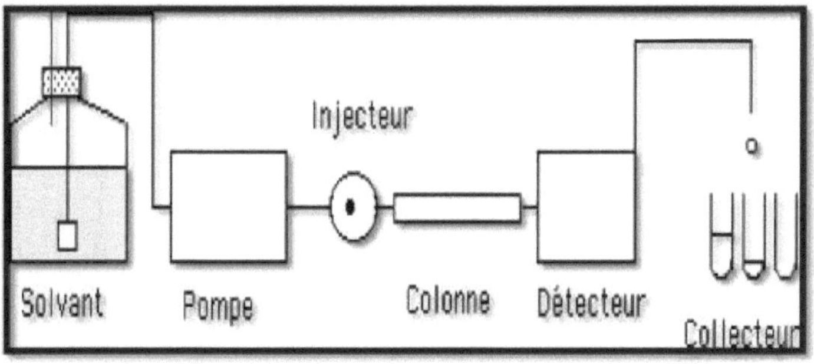

Figure 06: Liquid chromatography (LC - HPLC) [21]

8-2- Spectrometric techniques

Structural and quantitative analysis techniques.

8-2-1- Mass spectrometry

Analysis of the nature, composition and structure of the substance. Vacuum analysis poussë 10-4 Pascal. [21]

9- Preserving essential oils

Natural essences should be stored in the same way as the plants that contain them, in tightly closed containers away from light. They oxidise rapidly in light and air, polymerise and turn into resins, losing their odour and characteristics. [21]

10- Use of essential oils

The most common uses for essential oils are as follows:

- Self-medication for comfort, calm and relaxation (bath, massage, cosmetics)
- One of the components of traditional medicine and naturopathy[3].
- Aromatic use in psychology.

10-1-Medicine

- Disinfecting and healing wounds or treating trauma
- Complementing a chronic medical treatment.
- Dermatology and dermatological cosmetics
- But also the therapeutic choice made by a qualified aromatherapy therapist - doctor, pharmacist or naturopath. [3]

10-2-In industry

- Perfumery.
- Cosmetology.

10-3-Power supply

- Essential oils are widely used as food flavourings (fruit juices, pastries, especially hydrolats from aromatic plants) [3].

11-Examples of essential oils

11-1-Artemisia herba alba **essential oil**

The oil is obtained from the flowering tops or leaves. The essential oil is mainly composed of terpenes, but may also contain between 30 and 60% of cëtones. From a molecular point of view, it is composed of cham azulene, camphor, isothujone and limonene.

This oil has excellent anti-allergic, anti-inflammatory, antihistaminic and parasympaticolitic properties. Its choleretic, mucolytic and anticatarrhal properties are less good. This essential oil will suit you if you suffer from asthmatic or catarrhal bronchitis or asthma. The same

applies to hay fever, dermatitis, allergic eczema and urticaria.
Its action on those suffering from biliary insufficiency will not be as good as in the previous cases, but it will be no less remarkable. Note, however, that this oil is not recommended for beginners, as it is very powerful. It is also neurotoxic and abortifacient, so it should not be used on children or on pregnant or breast-feeding women. [45]

***11-2-Thymus vulgarus* essential oil**

An essential oil can be extracted from thyme leaves, and its properties are used in phytotherapy, particularly its antiseptic properties, which are used mainly in the lungs and digestive tract. These properties are due to the presence of linalool in thyme essential oil. Linalool is a phenol that is also found in the essential oils of several other plants with a characteristic odour. It is this phenol that can cause poisoning when consumed in excessive quantities.
Despite the fact that this plant is not toxic, precautions must be taken: thyme can be dangerous if the wrong dose is used. What's more, pollution and the way it is harvested can cause serious damage to the plant's virtues and to the plant itself. Thyme essential oil cannot be used pure; it must always be diluted. [46]

Chapter 3: Herbal medicine

1- History :

From the outset, man has cЬercЬё to soothe his ailments and reduce his suffering. To do this, he has ийНзё products immёdiately at his portae. The vёgёtal kingdom, which provided him with most of his food, was his first field of expёriences. Little by little, he learned to discern the properties, virtues and toxicity of plants. All the ancient civilisations - Mesopotamian, Egyptian, Chinese, Indian and Pre-Columbian - had an impressive array of plant remedies. Thus, over time, a relatively well-developed pharmacopёe [21] was built up.

For thousands of years, practical use was the only path to therapeutic progress, and all the knowledge acquired over the centuries, without any real the'onque approach or understanding of how plants work, constitutes the empirical data of tradition. [19]

The daily use of plants has led to the observation of a large number of their effects, which today's science recognises as very real. For example, the calming properties of opium from the poppy plant were known 4,000 years before we learned to extract morphine, which remains one of the major analgesics used in cancer research. All this knowledge, which for a long time remained empirical and has been made more rigorous by the progress of modern science [21], bears witness to the durability of phytotherapy.

In this way, mёdecine through plants is rich in lessons thanks to the inexhaustible wealth of observations accumulated century after century. What's more, the major pharmaceutical companies have made no mistake about it. In their quest for the new molecule, their research was directed towards vёgёtal substances or other products used by [19] traditional medicine.

Real clinical research has been possible for more than twenty years, thanks to the numerous studies carried out by the Sociёtё Frangaise de Phytotherapie et d'Aromatherapie. It has culminated in a new approach to medicinal plants, one that strives to redefine their medicinal use in a synthetic vision based on traditional data and modern pharmacological knowledge.

Recent developments in clinical phytotherapy therefore allow us to envisage three levels of approach in the use of medicinal plants: an empirical level, a pharmacological level and a clinical level. [21]

2- General

Traditional mёdications and the use of plants in empirical medicine have often been at the origin of high-level scientific research. In most cases, this research has led to the discovery of original substances of considerable therapeutic interest. These isolated molecules, purified and then used by conventional medicine, are widely available to industrial pharmacy; but plants were and are used successfully in phytotherapy and aromatherapy in various forms. These medications are increasingly popular with the public and with doctors. Plants cure or help to cure, sometimes very quickly, not only fatigue, insomnia, headaches, flu, coughs, rheumatism and colds, but also a great many other illnesses. [44]

3-Definition

Phytotherapy, from the Greek **Phuton** meaning plant and **Therapeia** meaning treatment, means treatment using plants.

Phytotherapy is the whole range of therapeutic treatments involving the direct use of drugs of plant origin. Plant materials can be used in their simplest form (simple or compound infusions) or in the form of galenic preparations (tinctures, extracts).

Phytotherapy represents a part of medicinal therapeutics, and is currently enjoying a veritable renaissance, both in the field of internal illnesses and in dermatology and cosmetology (soaps, deodorants, powders, etc.), as well as in baths and ear therapy (baths, compresses). [21]

Phytotherapy is an age-old practice based on empirical knowledge that has been passed down and enriched over countless generations. Phytotherapy, etymologically treatment by plants, is a therapeutic method that uses the action of medicinal plants.

There are two types of phytotherapy:

- A traditional practice, some of which goes back a very long way, based on the use of plants whose virtues have been discovered empirically. According to the WHO, phytotherapy is considered a traditional medicine and is still widely used in some countries.

including developing countries. It is a non-conventional medicine due to the absence of clinical studies. [38]

- A practice based on scientific advances which seeks out active plant extracts. The active extracts identified are standardised.

This practice leads to phyto-drugs and, depending on the regulations in force in the country, the circulation of phyto-drugs is subject to marketing authorisation (MA). This is known as pharmacognosy or pharmaceutical biology. [38] Phytotherapy is the art and science of medicine using plants, and is one of the oldest therapeutic methods.

Since ancient times, plants have been used to treat illnesses (in the form of powder, extract, tincture, infusion or decoction...), or to treat illnesses (through their seeds, roots, leaves, fruit, etc). [30]

4-The purpose of phytotherapy

The aim of phytotherapy is the same as that of any curative practice, which is to restore a deficient state of health. The advantage of phytotherapy is that it allows the doctor to establish a therapeutic strategy based on the possibilities offered by the plant and the objective set by the doctor. [5]

5-The principles of phytotherapy

Phytotherapy is based on two principles:

5-1-1ere Principle

It is preferable to use the plant's total rather than one of its isolated active ingredients. The total is defined as the most complete set of all the plant's active molecules. The total is an ordered set of substances produced by the plant's metabolism. [27]

5-2-2eme principle :

Introduce reactionary treatment rather than symptomatic treatment, because reactionary treatment does not replace the body in its fight against the disease, it stimulates the normal defence reactions that enable the body to return to a state of good health. [27]

6-Branches of phytotherapy

There are many phytotherapeutic practices:

6-1-Aromatherapy

This therapy uses only the essential oils of aromatic plants, such as thyme and menthol; sometimes toxic, these substances should be used [24] with care and in the prescribed doses.

6-2 -Gemmotherapy

This method is part of homeopathy. It uses alcoholic and glycerine extracts of young plant tissue (shoots, buds, etc.). [24]

6-3-Chinese phytotherapy

It belongs to the group of "traditional Chinese medicines". Like acupuncture. Its aim is to balance the body's energies or modify the circuit. [24] **6-4- Pharmaceutical phytotherapy**

It uses products of plant origin obtained by extraction and diluted in etiyl alcohol or another solvent. The dosage of these extracts depends on their action on the body.

These products can be used to prepare medicines comparable to those of any other pharmaceutical speciality (syrups, drops, capsules). [24]

7- Aromatherapy

7-1- History

In the history of medicine, at least until the 16th century, the history of aromatherapy is largely conflated with that of piytotierapy. Plants as a whole formed the basis of the piarmacopee of ancient civilisations. Although there are traces of distillation or extraction methods dating back several millennia in China and India, it was in Egypt that their use was first discovered.

In Greece, the writings of Dioscorides refer to the use of aromatic extracts. The Romans also used them in the form of oily ointments. In the X[e] century, the Arab alchemist Jabir Ibn Hayyan is credited with inventing the alembic. Extraction methods were subsequently improved, with pharmacopeias using them mainly after the 16th[e] century. It was not until the XIXe century that the active ingredients of odoriferous molecules began to be isolated and classified, enabling them to be used in specific ways. [21]

In the 1920s, the chemist Rene Maurice Gattefossë, who was researching perfumery, seriously burnt his hands. He dipped them in a container of lavender essential oil and was immediately soothed, then noticed that his burns healed quickly.

From then on, he devoted his research to the properties of essential oils. He coined the term aromatherapy, which has since become a household word. In the 1960s, Dr Jean Valnet took up Gattefosse's work and published reference works. They are considered to be the fathers of modern aromatherapy. Subsequently, Pierre Franchomme contributed to improving the quality of the extracts used, with the concept of the chemotype. (Encyclopedie Collaborative En Ligne, Universelle et Multilingue).

At the end of the XX[e] century, aromatherapy, like pharmacognosy as a whole, benefited from advances in analytical methods, particularly chromatography. The precise distinction between aromatic compounds has enabled medicine to gain a better understanding of their mechanisms of action, and to refine their prescription.

In the future, research into new active molecules is likely to involve [19] aromatic plant extracts once again.

Alongside these scientific advances, traditional aromatherapy is still very much a part of non-conventional medicine, or is used as a complement to treatment. It is still very popular with the general public.

Outside the West, aromatherapy and phytotherapy still play an important role in traditional medicine. [21]

7- 2- Definition

The birth of modern aromatherapy is due to the chemist Rene Maurice Gattefosse in the 1920's. Aromatherapy is a therapy using aromatherapy oils.

[21] from so-called "aromatic" plants.

Aromatherapy is the treatment of illnesses using plant aromas, i.e. aromatic essences known in medical parlance as essential oils.

This is a complementary practice to phytotherapy based on essential oils and concentrated oils, designed to physically and psychologically rebalance the individual. [23] **7-3- Aromatic biosynthesis**

The sun is the source of energy for life on earth. Without the energy of solar radiation, no form of life could exist on Earth. The beings that formed the first link in the life chain 3.5 billion years ago (cyanobacteria and blue-green algae). They drew their resources from the constituent elements of their universe: fire (light-energy), algae (water-energy) and water (water-energy).

solar radiation), air (oxygen and carbon dioxide), earth (minerals and trace elements), water (substances in solution).

Aromatic plant compounds (essences, balms) are by-products of photosynthesis, produced in specialised secretory cells located in flowers, fruit, ëcorces, wood leaves and roots....

These vëgëtaux qualified as aromatic, are used to obtain essential oils by steam distillation of the various parts of the vëgëtal. [17]

7- 4- Aromatic substances

This group includes a number of substances frequently found in plant drugs, which often vary greatly in composition and action. They may accompany other active substances in the plant, and this group includes phenolic glucosides [21] and phenylpropane derivatives such as coumarins.

7-5- Renewal of aromatherapy

Throughout history, traditional medicines based on natural substances have always proved effective.

Phytoaromatherapy is the oldest therapy in the world. It has always existed, since plants have never ceased to be used as condiments, medicines and ritual plants.

As opposed to drugs obtained by chemical synthesis or extraction of active ingredients, phyto-aromatherapy is natural medicine par excellence. The structure of an essential oil is complex and never possesses just one therapeutic property, but several. As a result, there are opportunities for synergy and potentiation, enabling the therapy to be tailored to individual needs.

But its main characteristic lies in the fact that it is a field medicine aimed at restoring the balance of an organism as a whole.

It's not a question of pitting "all chemical" against "all natural". Synthetic chemical substances have produced exceptional results that everyone recognises. These brilliant successes led to medicinal plants being forgotten for many years.

Essential oils, which often have a complex chemical composition, are powerful, reactive medications with a wealth of therapeutic properties that augur well for the future [44].

I . Aromatherapy is now gaining ground in the medical, cosmetic, agri-food and veterinary worlds. Confirmed scientific researchers from all over the world are determined to give this discipline its due by making great efforts in various areas of fundamental and applied research. [44]

Partie 02 :

Partie expérimentale

2 part

Practical part

Chapter 1: Materials and methods

1-Materials

1-1- Equipment

- Regulated magnetic stirrer
- Bain marie
- Magnetic strip
- Wood lamp
- The high
- Hydro-distiller

1-2-Tools and glassware

- Platinum shackle
- 200-1000 ml beakers
- Funnel
- Swabs
- Sterile discs
- Para-film
- Aluminium foil
- Toilet paper
- Pliers
- Pipette

1-3- Products

- Agar- agar
- Ethyl acëtate
- Distilled water
- Autoclave
- Precision scale

-Bunsen burner

-Vortex

- Regulated oven at 37° C
- Digital cameras

- Test tube

- Sterile plastic Petri dishes
- Glass bottles
- Sterile tips
- Micropipette
- Cloth
- Marker pen
- Blotting paper
- Spatula
- Vice tubes
- Bleach (antiseptic)
- Sterile physiological water
- Ethanol
- Hexane
- Commercial oils: Thyme and Mugwort

- Essential oils from two species: *Thymus vulgarus* and *Artimisia herba alba*
- Sabouraud Dextrose AGAR
- Mueller-Hinton Bouillon.

1-4-Biological equipment

1-4-1-Yeasts

- EspeceOl: ***Candida zeylano'ides***
- Species 02: ***Candida parapsilosis***
- Species 03: ***Candida albicans***
- The 03 species are isolated from honey (institut national de science mëdicale. Universite B Elaji Mokhtar (Annaba).
- **1-4-2-Mould**
- *Aspergillus sp* (petrified brood) isolated from a bee larva in a willaya hive at El-Taref

2-Methods

This study, which was carried out in the microbiology laboratory of the Mohamed Cherif Messaadia University in Souk-Ahras, aims to identify the effects of essential oils on 4 species of fungi.

2-1-Growing environments :

2-1-1-Middle Sabouraud :

Naturally acidic, it inhibits the growth of many bacteria. Sabouraud's gelose is a classic medium for culturing, isolating and identifying saprophytic or pathogenic yeasts and moulds. It is mainly recommended for isolating moulds in samples with low bacterial load, and for sterility testing of pharmaceutical, cosmetic or food products.

In the case of heavily contaminated samples, it is preferable to use Sabouraud gelose + chloramphenicol. [51]

***J* Preparing the environment**

To prepare this environment, follow the steps below:

- Place 1000 ml of distilled water in a Bucher with a stirring rod.
- Weigh 61.4 g of Sabouraud Dextrose AGAR using a precision balance and place this quantity in the previous bucher;
- Shake the mixture with a standard magnetic stirrer to obtain a homogeneous mixture;
- Use a funnel to pour the mixture into glass bottles;
- Sterilise the medium in an autoclave at 120°C for 20 minutes.

2-1-2- Mueller-Hinton midfield

Non sëlective base medium for bacterial culture (antibiogram), it is a relatively rich medium, but remains a base medium that allows the culture of non-demanding bacteria.

Mueller Hinton gelose is a solid medium used for antibiotic susceptibility testing of pathogenic bacteria. This medium is used routinely for sensitivity testing. [51]

***J* Preparing the environment**

To prepare this medium, follow the steps below:

- Introduce 21g of Mueller-Hinton Bouillon with 1000 ml of distilled water into a beaker with a stirring rod;
- Add 20g of agar agar to solidify the medium;
- Stir the mixture with a standard magnetic stirrer until a homogeneous mixture is

obtained;

- Use a funnel to pour the mixture into glass bottles;
- The medium is sterilised in a controlled autoclave at 120°C for 20 minutes.

2-2 - Quantitative and qualitative analysis of essential oils

2-2-1- Quantitative analysis

- Yield: the essential oil is obtained from 25 g of dry matter.

A-Dry matter :

The aim of this step is to see the water content of these two plants, because we have used them undried.

- Weigh out 5g of thyme and 5g of mugwort
- Place in a 105°C oven for 24 hours.
- Weigh the thyme and mugwort again and see the difference in weight.

B-Extraction of essential oils :

Hydro-distillation and steam distillation are extraction methods approved by the French pharmacopoeia, as well as by AFNOR (the French standards association).

In our study we used the steam stripping extraction method because it has several advantages, such as: the speed of the operation (2h,

30min), obtaining better yields and not altering fragile essential oils.

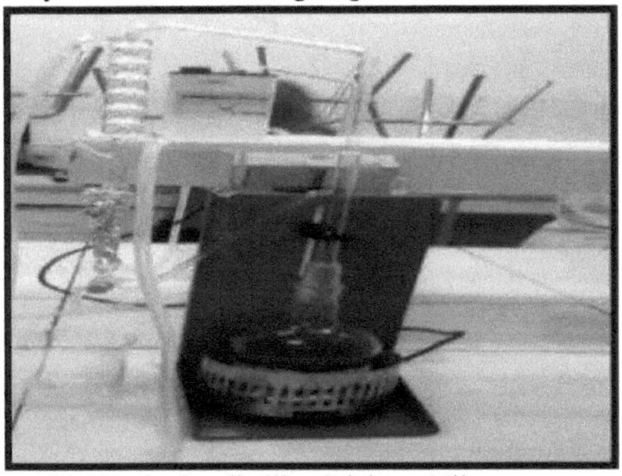

Figure 07: The hydro distiller (Lickens Nikerson apparatus)
(Mohamed Cherif Messaadia University, 2013)

C- Principle

Vapour entrainment consists of separating the volatile products contained in the vëgëtal drugs (***Thymus vulgarus*** from the Souk-Ahras wilaya, ***Artemisia herba*** **alba** from the **Boukhadra** wilaya region of **Tebessa**) which are subjected to contact with the vapour. The water saturated with essential oils passes through a coil where it condenses to give two products, floral water and the essential oil at the outlet (the oil is separated from the water by its density). **D- Technique used**

The vëgëtale (the leaves of the plant) prepared later is put in a 500ml flask which is filled with 99ml of distilled water, the flask with its contents will be put on the balloon heater. The EOs are entrained by the water vapour generated in the flask and directed towards the swan neck (elbow) which connects the flask to the refrigerant (water source).

Once in the refrigerator, they condense rapidly and precipitate in the decantation ampoule, which allows immediate separation according to density: floral water and essential oil, which will be recovered by eliminating the water through the opening of the ampoule. After 2 : 30 hours of extraction, the essential oil is collected in a small hermetically sealed glass bottle, covered with aluminium foil to protect it from light and prevent oxidation, and stored in the refrigerator.

2-2-2- Qualitative analysis

> Identifying the different components of essential oils

A- Identification of the various components of essential oils by TLC

There are various methods for identifying essential oils, including: "Thin layer chromatography (TLC): This is a method for separating the chemical groups of essential oils.

8- The principle

- Chromatography is based on the differential migration of the various components and their distribution between the mobile phase and the stationary phase. In thin layer chromatography, the stationary phase is deposited in a thick film (1mm-2.5mm) on a solid surface, either rigid (glass) or flexible (plastic or aluminium), which is easy to cut.
- The solution of the mixture is initially deposited as a dot on the plate and introduced into a sealed vessel containing the diluent (mobile phase).
- The constituents of the mixture are separated as the mobile phase rises by capillary action along the stationary phase.
- Colourless substances will be made visible (detection using a developer).

C- Operating procedure :

The aluminium oxide chromatography plates are placed in the oven at a temperature of 105°C for 15 minutes.

Solutions of the essential oil diluted in ethanol (add 0.2^L of ethanol to 0.02^L of essential oil) are deposited on an aluminium oxide chromatography plate to separate the chemical groups by means of a regular spot 1cm from the bottom edge of the plate for each oil.

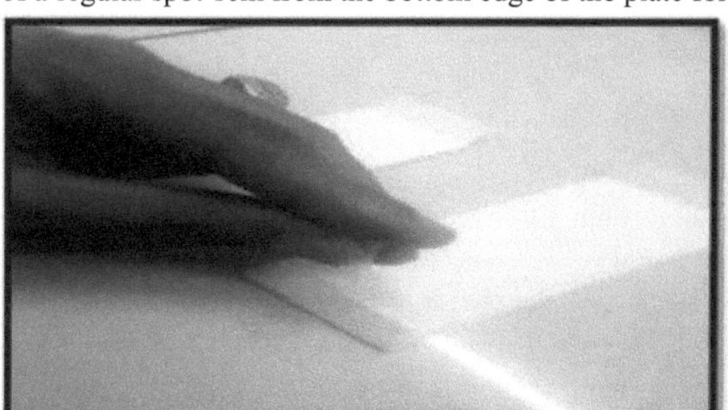

Figure 08: Arrangement of EO (thyme and mugwort) on TLC plates **(Universite Mohamed Cherif Messaadia, 2013)**

- The solvent is prepared: 96% hexane + 4% ethyl acëtate
- 28.8 ml of hexane and 1.2 ml of ethyl acëtate are added to 30 ml of solvent.
- The 2 plates are placed in 2 beakers each containing 15 ml of solvent after the diluted oils have been dried.

When the solvent and the substance to be chromatographed are placed above the level of the mobile phases, the chambers are closed and kept at a temperature of 20° to 25° C. The 2 plates are removed as soon as the solvent has reached a known distance, then air-dried.

Figure 09: TLC plates deposited on a mobile phase (solvent)

(Mohamed Cherif Messaadia University, 2013)

D-Identification of separate substances

A Wood's lamp (UV254-365) is used to reveal the spots, then each spot is located and the frontal ratios calculated.

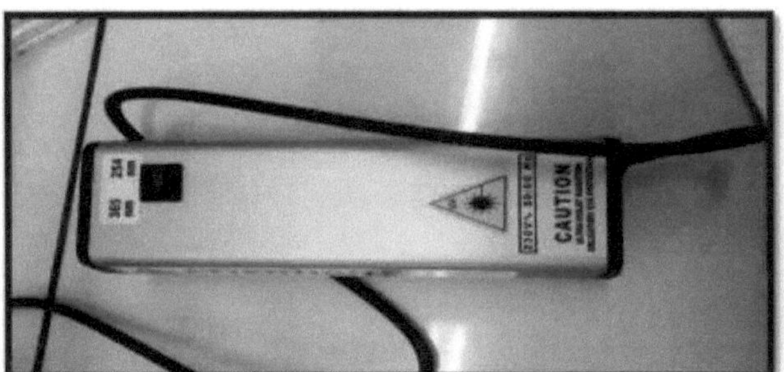

Figure 10: The Wood lamp

(Mohamed Cherif Messaadia University, 2013)

> **Face ratio value in TLC:** The arrangement of substances on chromatography is exprптёе most often in face ratios (FR) or face ratio height (HRF).

Calculation of the frontal ratio: To determine the frontal ratios of the constituents of essential oils, these stains were detected by Wood's lamp using UV rays. „"

$$RF = \frac{\text{Distance entre le centre de la tache et le point de départ}}{\text{Distance parcourue par la phase mobile}}$$

Distance between the centre of the spot and the starting point

Distance travelled by the mobile phase

Calculating the face-ratio height: The face-ratio height is the product of the RF X100.

HRF = RF X 100

2-3-Sowing and incubation

2-3-1-Sowing on Sabouraud

The 03 yeast species: ***Candida zeylano'ides; Candida parapsilosis; Candida albicans*** and the same as for the mould species: ***Aspergillus sp*** isotees are streaked in Petri dishes containing Sabouraud medium 4mm thick after solidification and using a platinum loop.

Incubation takes place at 37°C for 5 to 7 days.

2-3-2-Sowing on Mueller-Hinton

Seeding

-I- Stage 01

After the growth of yeasts in Pë^Б dishes containing Sabouraud medium, these yeasts rëensemencëes in Pëtries dishes containing Mueller-Hinton as follows:

- Pick growing yeast colonies on Sabouraud medium by a ëstetrile swab for each species, and placing it in the tube containing sterile physiological water (one tube for each species);
- Wring it out by pressing it firmly (by turning it) against the inner wall of the tube, in order to unload it as much as possible to obtain the yeast suspension ;
- Stirring the mixture with a vortex;
- Rub the swab over the whole of the dry gelose surface, from top to bottom, in streaks;
- Finish seeding by passing the swab around the edge of the gelose.
- If you seed several Petri dishes, you need to reload the swab each time.
- The stëriles blotting paper discs are placed on the yeast-containing surface;

- Different concentrations of essential oils (*Thymus vulgarus, Artimisia herba alba*) are added, starting at **4ul; 6ul; 8ul; 10 ul; 20ul; 40ul; 60ul; 80ul** for the 04 species of fungi ***(Candida zeylano'ides; Candida parapsilosis; Candida albicans; Aspergillus sp).***

NB: there are two repeats for each dose and a control.

Incubation: in the incubator at a temperature of 37°C for 5 to 7 days.

-I- Stage02

The aim of this step is to see whether the spores adjacent to the zone of inhibition in the Petri dishes in which the wormwood essential oil was used cringe or not.

Seeding

- With the aid of a platinum loop, quantities of the 04 species of fungi (*Candida zeylanoides; Candida parapsilosis; Candida albicans; Aspergillus sp)* adjacent to the inhibition zone in which Armoise essential oil was used were prepared; starting with doses of **4ul, 6ul, 8ul and 10ul.**
- Spread the quantities taken over the dry Sabouraud gelose in streaks.

Incubation: in a incubator at 37°C for 5 to 7 days.

A Petrie dish is used for each dose for a specific species.

-I- Step03:

This step is to see how effective the commercial oils (Mugwort and Thyme) are on the species of fungi, and to compare the effects of these oils with the effects of the essential oils (Mugwort and Thyme).

Seeding:

- Collect colonies of 04 species of fungi ***(Candida zeylano'ides; Candida parapsilosis; Candida albicans; Aspergillus sp)*** using a sterile swab for each species;
- Rub the swab over the whole of the dry gelose surface, from top to bottom, in streaks;
- Finish seeding by passing the swab around the edge of the jelly.
- If you seed several Petri dishes, you need to reload the swab each time.
- Place the sterile discs on the yeast-containing frozen surface

- Different concentrations of commercially-available oils (Thyme, Mugwort) are added, with concentrations of 80цl; 100цl for the 04 species of fungi ***(Candida zeylano'ides; Candida parapsilosis; Candida albicans; Aspergillus sp*).**

Incubation: in the incubator at 37°C for 5 to 7 days.

NB: All the inoculation steps are carried out in a sterile area around the Bunsen burner with a diameter of 15 to 20 cm.

Chapter 2: Results and discussion

1-Results

1-1-Dry matter :

Table 01: Dry matter weight of the two plants

The plant	Before drying	After drying
Artimisia herba alba	5g	4,4g
Thymus vulgarus	5g	4,6g

1-2- Essential oil yield

The yield of essential oils from 25g of dry matter for each plant species (*Artimisia herba alba, Thymus vulgarus*) is shown in the table below:

Table 02: Yield of essential oils from two dëfërent plants.

Essential oils	Dry matter weight	Volume (ml)
Artimisia herba alba	25g	**0.27ml**
Thymus vulgarus	**25g**	**0.66ml**

1-2-1- Organoleptic study of essential oils

Once the essential oils had been collected, they were ël.ë preserved in hermetically sealed glass bottles protected from light, and their organoleptic characteristics were determined.

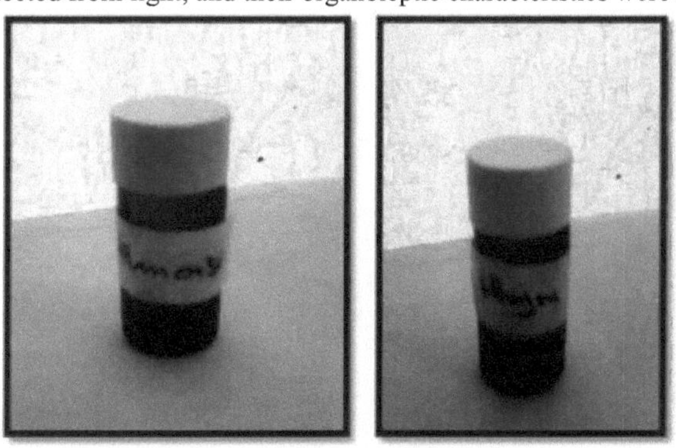

Figure 11: Storage of essential oils in airtight glass bottles

> Mugwort essential oil (*Artimisia herba alba) :*

Hi **Appearance**: liquid, clear

Hi **Colour**: pale yellow

Hi **Odour**: very aromatic

Hi **Flavour**: pleasant and bitter

> Thyme (*Thymus vulgarus)* essential oil:

Hi **Appearance**: liquid, clear

Hi **Colour**: pale yellow

Hi **Odour**: very aromatic

Hi **Flavour**: pleasant and bitter

1-2-2- Characterisation of essential oils by TLC

By separating the constituents of the essential oils using thin layer chromatography, we were able to determine the difference between the chemical composition of the essential oils of the two different oils.

The frontal ratios (FR) found were compared with bibliographic data proposed by STAHL. (1975).

HRF= (Distance travelled by the spot/ Distance travelled by the mobile phase) x 100

A /Mugwort essential oil (*Artimisia herba alba)*

For the mobile phase **(ethyl hexane-acetate)** we observe spots on the 1ere TLC plate using the Wood lamp reveals the existence of two spots with different Rf.

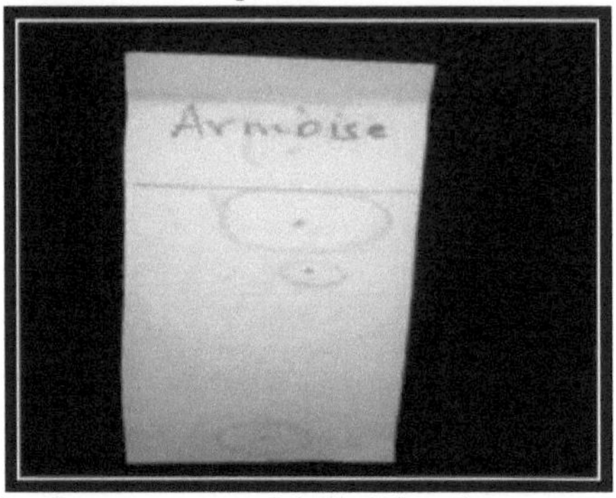

Figure 12: TLC analysis of mugwort essential oils (Plate 1)

Table 03: Rësults of TLC analysis of Armoise HE (*Artimisia herba alba)* mobile phase.

Stains	Rf	HRF
01	0.62	62
02	0.77	77

NB = The spots are numërotëes from bottom to top

- The results of TLC analysis of Armoise essential oils (*Artimisia herba alba*) show the presence of two chemical groups.

B /Thyme essential oil (*Thymus vulgarus)*

For the mobile phase **(ethyl hexane-acetate)** we observe spots on the 2eme plate using Wood's lamp ^ëк the existence of two spots with different Rf.

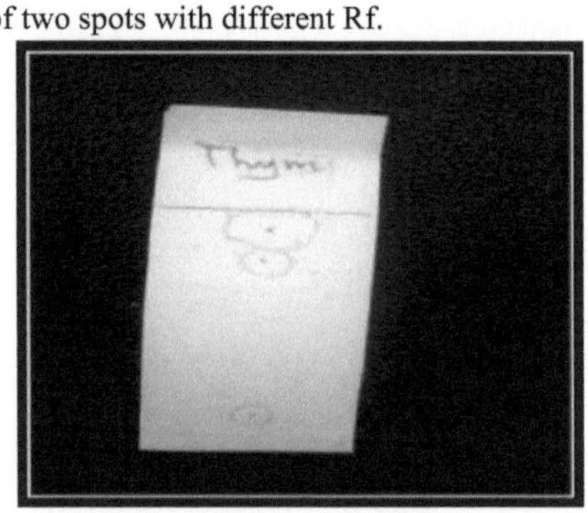

Figure 13: TLC analysis of thyme essential oils (Plate2)

Table 04: Results of TLC analysis of thyme (*Thymus vulgarus)* EO for the mobile phase.

Stains	Rf	HRF
01	0.75	75%
02	0.91	91%

NB = Spots are numbered from bottom to top

- The results of TLC analysis of Armoise essential oils (*Artimisia herba alba)* show the presence of two chemical groups.

Interpretation of antifungal activity test results

The results of the aromatogram show real activity for all the samples testedës. Nevertheless, this activity differed according to the type of essential oils, concentration and type of target fungi.

Table 05: The antifungal effect of essential oil (*Artimisia herba alba*) on different fungi at different doses.

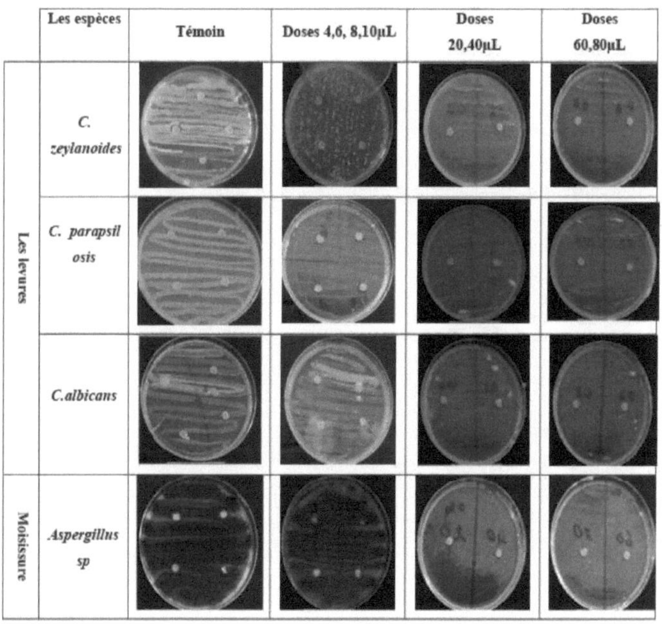

Table 06: Diameters of the zones of inhibition of mugwort oil on fungi.

Fungal strains	0 of the muting zone (mm)				Note
	Dose 4	Dose 6	Dose 8	Dose 10	
C.zeylanoides	9	10	11	8	++
C. parapsilosis	10	12	10	12	++

C .albicans	8	10	12	14	+++
Aspergillus sp	0	0	0	0	-

- For ***C. zeylanoides, the*** activity of mugwort essential oil on this species is less weak compared to tëmoin with colonies sëparëes for low doses (4, 6, 8,10ц1) at the range of variation of the diameters of the zones of inhibition varies from **[8 - 10mm]** but the antifungal activity of the latter was increased with the increase in the concentration of essential oil.
- For ***C. parapsilosis*** there was a weak growth of fungi with antifungal activities at the range of variation of the diameters of the zones of inhibition varies from **[10-12mm]** and the fungal action was increased with the increase in the concentration of essential oil.
- For ***C .albicans*** there was antifungal activity with the range of variation of the diameters of the zones of inhibition varying from **[8 -14mm]** and the antifungal activity was increased with the increase in the concentration of essential oil.
- For ***Aspergillus sp*** there is a good growth of fungi with low concentration of essential oil is low but there is a great antifungal activity in the concentration of **80 ill of** essential oils that qualifies as excellent action.

Determination of the minimum inhibitory concentration for 1 pure essential oil (*Artimisia herba alba*):

Table 07: Tests for the determination of the minimum inhibitory concentration for pure essential oil (*Artimisia herba alba*).

	Les Espèces	Zone de dose 4 µL	Zone de dose 6 µL	Zone de dose 8 µL	Zone de dose 10 µL
Les levures	*C. zeylanoides*	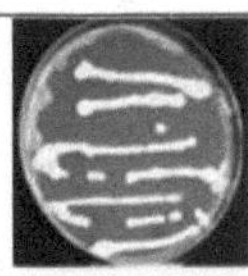	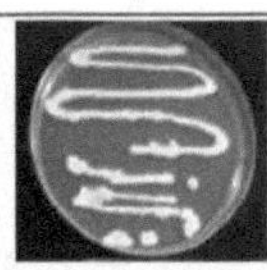	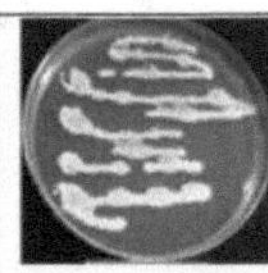	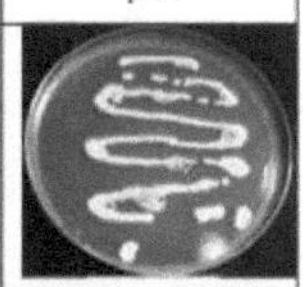
	C. parapsilosis			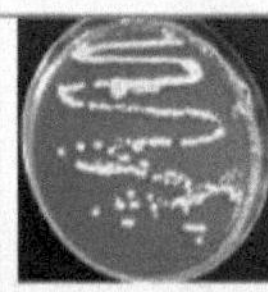	
	C.albicans	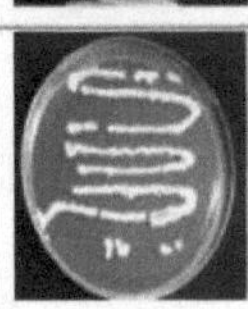	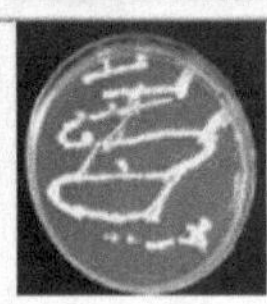	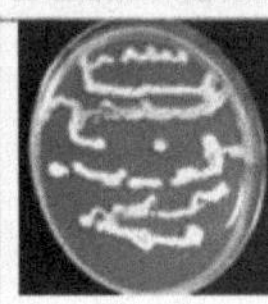	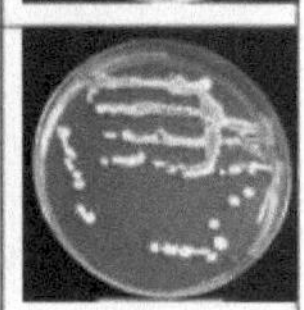

The table above shows the minimum inhibitory concentration after treatment with mugwort essential oil. Good yeast growth was observed with all 4 doses.
According to the latter, the effect of mugwort in doses 4, 6, 8 and lOgl is fungistatic because it only causes inhibition.

The commercial oil effect (Mugwort)

Table 08: Effect of commercial oil (wormwood) on various fungi

	Les souches	Témoin	Doses 80 µl, 100µl
Les levures	*C. zeylanoides*		
	C. parapsilosis	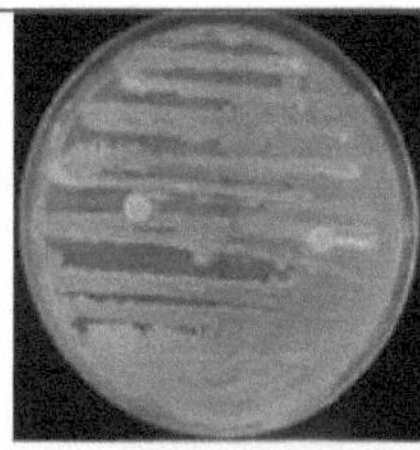	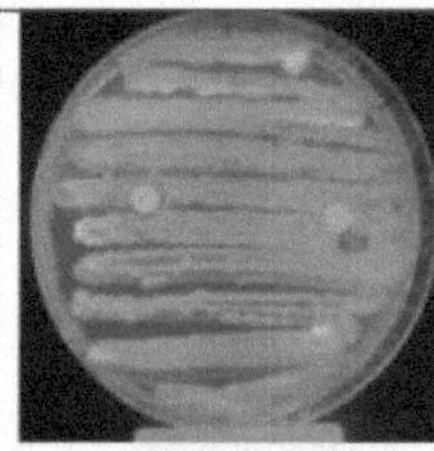
	C.albicans	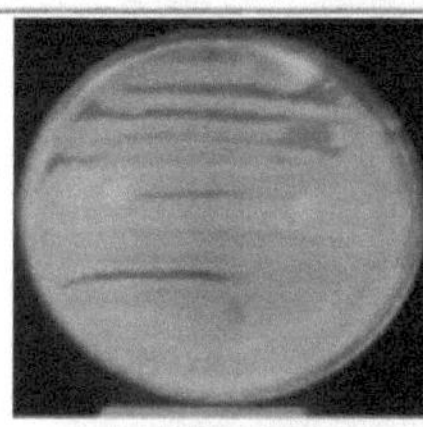	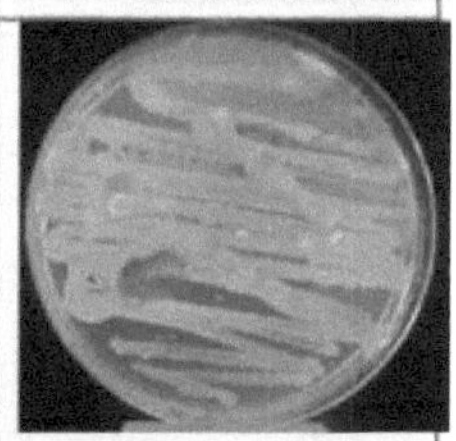
Moisissure	*Aspergillus sp*	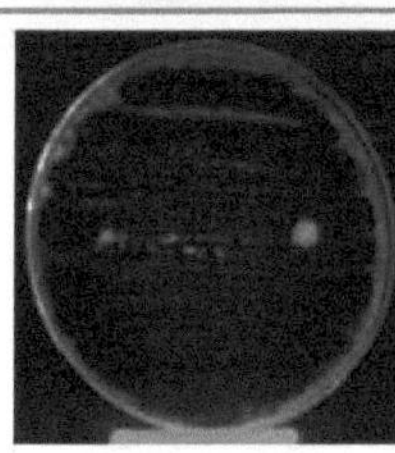	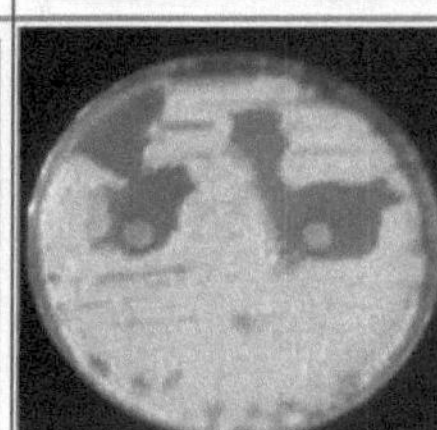

Interpretation: Using commercial mugwort oil, we saw no effect on the 03 species of yeast where growth was excellent and good with the high doses (80 and 100ɥl) but *Aspergillus sp*, when compared with the control we noticed that there was a delay in growth (white colonies) instead of black with zones of inhibition (15mm) for the 80ɥl dose and (17mm) for the 100gl dose.

Thyme essential oil:

Table 09: The effect of essential oil (*Thymus vulgarus*) on different fungi at different doses.

	Les espèces	Témoin	Doses 4,6, 8,10µL
Les levures	*C. zeylanoides*	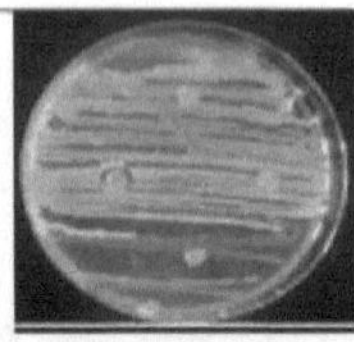	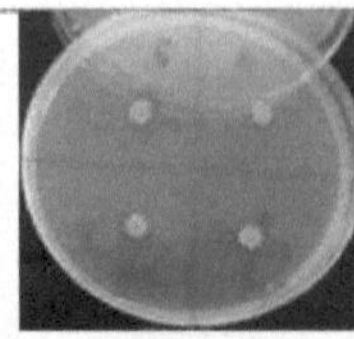
	C. parapsilosis	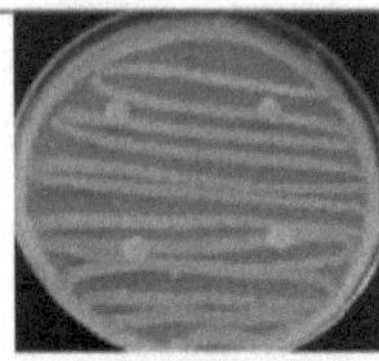	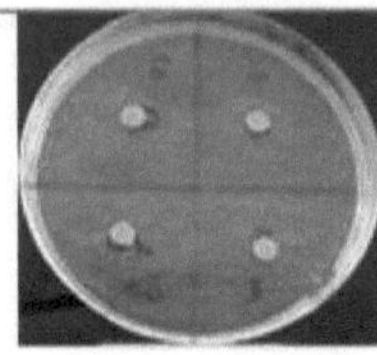
	C. albicans	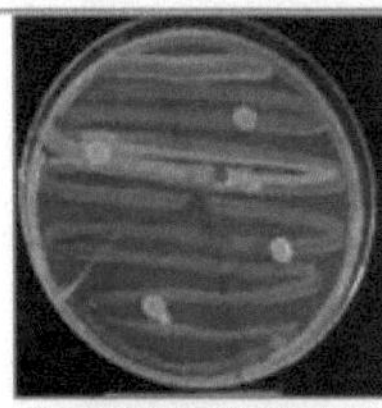	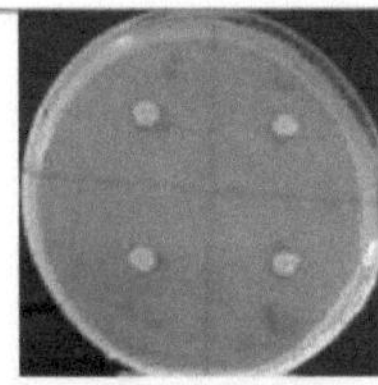
Moisissure	*Aspergillus sp*	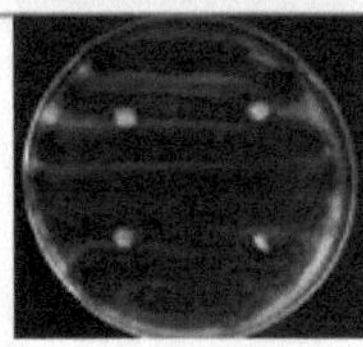	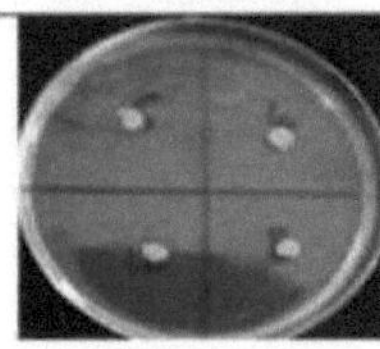

Interpretation: After using thyme EO in doses of 4, 6, 8 and 10pl, there was a total absence of growth and multiplication of yeast colonies and of the mould *Aspergillus sp* when compared with treated controls.

We can say that thyme has a real (total) fungicidal effect because it causes the colonies to disappear completely.

The commercial oil effect (Thyme) :

- For ***Candida zeylanoides*** there is a weak growth which explains a good antifungal action.
- For ***Candida parapsilosis and C.albicans,*** no antifungal action caused by the oil.
- For ***Aspergillus sp***, there is antifungal activity with low growth rates in the early stages.
- The following table shows their effect on fungi:

Table 10: Effect of commercial oil (Thyme) on various fungi

	Les espèces	Témoin	Doses 80 µl, 100µl
Les levures	*C. zeylanoides*	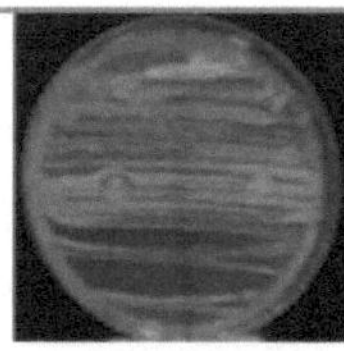	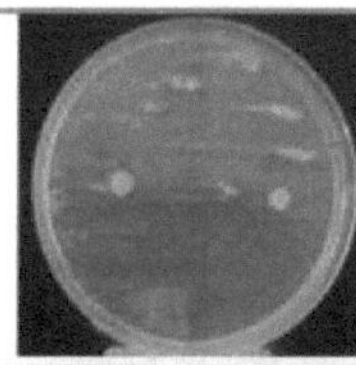
	C. parapsilosis	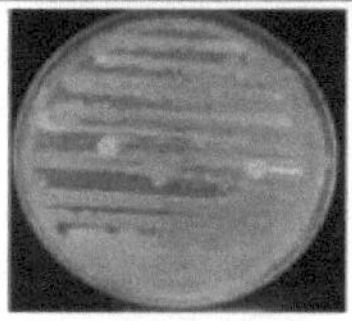	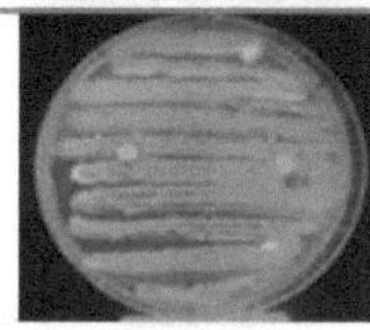
	C .albicans	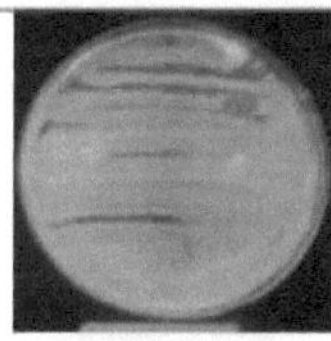	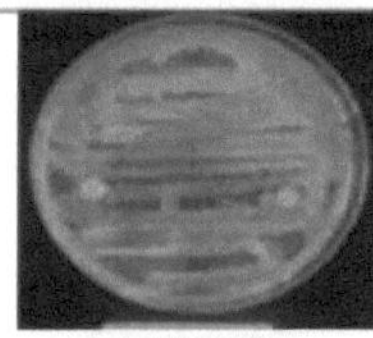
Moisissure	*Aspergillus sp*	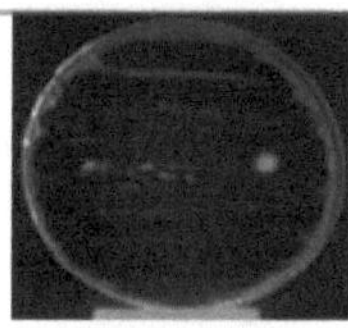	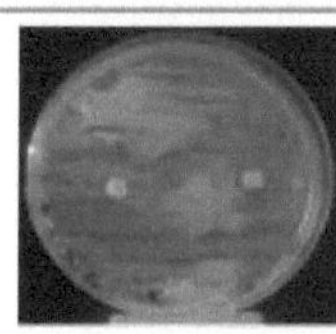

Interpretation: Using commercial thyme EO, we noticed that it had an effect on the two species *C. zeylanoides* and *Aspergillus sp*, where colonies showed little growth, but no effect on *C. parapsilosis* and *C. albicans* and no antifungal activity with commercial thyme EO.

The effect of mixing commercial oils (mugwort and thyme) on *Aspergillus sp.*

Good antifungal activity was observed on the *Aspergillus sp* mould species when two commercial oils **(wormwood and thyme)** were mixed at different concentrations **(80 ,100)** and the range of variation in the diameters of the zones of inhibition varied from **[14mm - 35mm].**

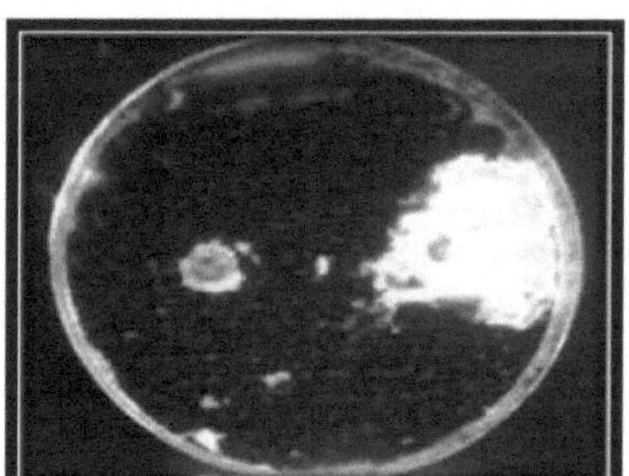

Figure14: The effect of mixing commercial oils (mugwort and thyme) on *Aspergillus sp.*

Interpretation : After the sensitivity of ***Aspergillus sp*** towards the commercial EOs and as a test (an idëe realiser) we mëlangë the two commercial EOs (thyme and mugwort) with doses 80, 100pl. We see a good effect on *Aspergillus with* two zones of inhibition of 14mm for 80цl and 35mm for 100pl.

2-Discussion

In view of the increased use of antibiotics in recent years and the growing resistance of micro-organisms (bacteria, fungi), we need to look at new natural substances such as essential oils as alternatives to antibiotics.

Our study was conducted with the aim of demonstrating the efficacy of two essential oils (thyme and mugwort) against 4 species of fungi pathogenic to humans and animals, such as *Aspergillus sp*, *Candida zeylapanoides, C. parapsilosis* and *C. albicans.*

In our laboratory, we extracted two essential oils using the hydrodestillation method because, according to **Abdelouahid et *al* (2010) [1]**, 'this is the method that keeps the aroma and fragrance components naturally present in the plant intact.

The average yield of essential oil was calculated on the basis of the dry plant matter in the aerial part of the plant (the leaves). The yield of essential oil from *Artemisia herba alba* was 0.27ml and the yield of essential oil from *Thymus vulgarus* was 0.66ml because the water content of *Thymus vulgarus* is higher than that of *Artemisia herba alba*, the type of plant and the environmental conditions.

Using the TLC method, we found that each of the two extracted EOs is made up of two chemical groups.

Mugwort and thyme essential oils show interesting antifungal activity in relation to the concentrations used, as highly pathogenic *Candida* and *Aspergillus sp* strains are highly sensitive to these two oils.

All the strains tested were extremely sensitive to all the doses of essential oil of thyme tested, showing a high sensitivity to lower doses of volume 4 pl. According to **Feerman et *al (*2006*)* [22]**, this can be explained by the great effect of thyme components on the ultrastructure of *Aspergillus sp*, where they cause irreversible damage to the plasma membrane and organelles of the fungus, in particular the mitochondria, which stops respiration and causes death, while inhibiting germination of the fungi.

spores, mycelium elongation, sporulation and toxin production in moulds, thyme oil causes lysis of yeast cells or it reacts with membrane enzymes and degrades the cell membrane and causes leakage of cell contents to the outside, and according to **Suppakul et *al.(* 2003) [41]** have suggestedërë that the antifungal activity of HE, can take place according to two different mëchanisms: some constituents cause the leakage of ëlectrolytes and the depletion of amino acids and sugars, others can be inserted into membrane lipids, consequently there y loss of membrane functions.

Commercial thyme has been shown to be less effective even at high doses (80, 100pl), where differences in the sensitivity of the species studied have been noted. This can be explained by the addition of products such as tween 20, 40, 60 and 80, which change their structure.

On the other hand, pure mugwort essential oil is less effective than thyme essential oil, with the volumes tested showing that mugwort in low doses is not fungicidal (letal) but is fungistatic (inhibitory), the proof being the multiplication of isolated spores around the inhibition zones, but the latter shows strong efficacy at volume 80 pl. This activity depends on the concentration of the oil, which shows good antifungal power. The fungi showed increased sensitivity to increasing the volume of oil in their culture medium where the colony diamëtre reduced each time the dose of oil was increased until total inhibition or no growth was observed, this may be due to the presence of more active compounds with increasing dose of essential oil.

Commercial wormwood is remarkably ineffective against yeast species, but remarkably effective against mould species.

We tried to see the effect of mixing two commercial oils (thyme and mugwort) but only on *Aspergillus sp did* we see an antifungal effect with the higher dose.

According to[22], , thyme essential oil is more effective than mugwort, and its antifungal activity explains why the molecular groups with the most powerful antibacterial actions are also effective antifungals. The results show that oils with strong antifungal properties, such as thyme and cinnamon, could be an interesting alternative in antimycotic therapies.

According to [35], thyme oil is often reported to be one of the most active oils. Its

Carvacrol, the main component, also has strong antimicrobial activity. These same authors pointed out that the activity of thyme oil is not due to the activity of a single component alone, but to the synergy between the other constituents.

As the majority constituent, thymol could be considered responsible for the strong antifungal power of thyme oil. This has been confirmed by the study of [15], this compound inhibited the development of pathogenic and phytopathogenic fungi at very low MICs. Thymol and carvacrol proved to be the most effective, followed by a-terpineol and the synergistic effect between salicylic a^hyde and linalool at a proportion that resulted in total germ inhibition.

Thanks to the richness of its main compounds, the essential oil proved effective against the fungal strains tested. These preliminary results can be supplemented by other, more in-depth studies (toxicological tests, method of application, etc.) in order to exploit their antimicrobial properties.

Conclusion

Most plants are useful to man. Thyme and mugwort are plants that have long been known in Alguriex and have traditionally been used as antiseptics, against aches and pains, for the hair,... .

Essential oils and their constituents have a long history as antifungal agents. The aim of this work is to make a contribution to establishing the antifungal activity of essential oils.

Initially, this study focuses on the extraction of these metabolites from native plants *(Artimisia herba alba, Thymus vulgarus*), using the hydrodistillation process, which yields

The HE iso^ were analysed by thin layer chromatography (TLC) to determine their biochemical composition. Significant differences were found in the quantitative and qualitative composition of the extracts obtained.

Secondly, the work is directed towards demonstrating the antifungal activity of these oils on pathogenic fungi (*Candida zeylanoides; Candida parapsilosis; Candida albicans and Aspergillus sp).*

The thyme EO studied showed strong antifungal activity on all the strains tested, while the mugwort essential oil showed average antifungal activity when compared with thyme.

However, it should be pointed out that the antifungal activity of thyme EO is not only due to the majority composës but to all the basic composëes.

The aromatogram of commercial mugwort oil showed efficacy on *Aspergillus*, while thyme oil showed antifungal activity on *Candida zeylanoides* and *Aspergillus.*

These results point to a favourable outcome for health safety.

References

[1]-Abdelouahid.D and Bekhechi. C., (2010) Essential oils. Edition © office des publications universitaires. 55p.

[2]-Abdelkader. B., (2011) Les plantes medicinales d'Algerie.

[3]-Amor. M., (2006). Phytotherapie, recolte des plantes medicinales, les huiles essentielles.

[4]-Anonymous, (2007). Medicinal plants. Centre for Agricultural and Rural Cooperation Postbus 380,6700 AJ Wageningen, The Netherlands. PP1-3.

[5]-Belaiche. P., (1979). Traite de piytotierapie et aromatierapie, tom 2, maloine S.A. Editeur paris.

[6]- Ben Mezrabtia .A et Difi .D. Caracterisation ciimique des iuiles essentielles et leurs activites antibacterienne de deux espëces de Lamiaceae *Origanum vulgarus* et le *Thymus vulgarus* dans la region d'Ouled Driss (Souk Ahras). Souk Ahras University Edition. P20.

[7]-Bego. Ph., (2001). Connaitre l'essentiel sur les huiles essentielles. Collection aromatherapie pratique, overview of various methods of extracting essential oils. Infoessence. PP5-6.

[8]-Boubrit .S., (2007) Determination "in vitro" of the antibacterial power of essential oils of eucalyptus, myrtle, cloves and savory, and their application to the preservation of fresh meat type hachee. Diploma of Ingeniorat d'etat in biology. Mouloud Mammeri University, Tizi-Ouzou .P59

[9]-Bourafa .I., (2007). Ethnobotanical study and control of common oregano. Memoire de fin d'etudes en vue de l'obtention du diplôme de pharmacien d'etat. Published by the University of Annaba. P14.

[10]- Chefrour. M and Hamaida. H.,(2005). Etude ethnobotanique des plantes medicinales dans l'Est Algerien (region de Souk-Ahras), Memoire de fin d'etude en vue de l'obtention du diplôme de pharmacien d'etat,partement de pharmacie, faculte de medecine, universite Badji Mokhtar, Annaba.PP34-36 .

[11] - Chopra I., Abrol B. and Handa. K.,(1960) Medicinal plants of arid regions considered especially from the botanical point of view. Drug Research Laboratory, Jammu (India).99p. **[12]-Clement. , (1981).** Larousse agricole. Edition : Paris. PP: 730-732.

[13]- Clevenger JF, (1928). "Apparatus for volatile oil determination, Description of New Type". American Parfumer & Essential Oil Revie.PP 467-503.

[14]-Cosentino. S and Tuberoso. C.,(1999). in vitro antimicrobial activity and chemical composition of Sardinian thymus essential oils. P 29.

[15]-Couladis. M., Tzakou. O., Kujundzic. S., Sokovic .M. and Mimica-Dukic N., (2004). Chemicals analysis and fungal activity of *Thymus striatus*. Phytother. Res, PP 40-42.

[16]-Danoune. K et Dziri .F. ,(2008). Etude de l'effet des huiles essentielles de Menthe et Thym et de leurs interactions sur certaines bacteries pathogenes de la viande. These d'ingënieur. Universite Mouloud Mammeri. Tizi Ouzu. 103p.

[17]-Dewit H., (1986). Plants of the world: flowering plants. Tome 2. Edition : Hchette. Paris. PP196- 201.

[18]- Djeroummi. A et Nacef. M. , (2004). 100 plantes medicinales d'Algerie, PP88 -159 .

[19] - Donadieu. Y. ,(2008). History of phytotherapy.

[20]-El Abed. D and Kambouche. N. (2003) Les huiles essentielles. Edition El Gharb. Oran.

61p.
[21]-Fellah. I., (2010). Etude d'une plante aromatique et medicinale Lavandula stoechas L.dans la ParcNational d'El Kala, Nord-Est Algerien ; Aspect Phytochimique, Histologique, et Antib acterien. Memo ire de l'obtention du diplôme de Magistere. Universite Badji Mokhtar, Annaba .102p.
[22]-Freeman. L and Carly. ,(2006). Aromatherapy: Nature news sience, nitration, prevention and scent. 56p
[23]-Fronty and Laura, (1995). Les plantes aromatiques, les connaitre, les cultivent, les utiliser. Edition: du chene. Paris. 96p.
[24]- Georges. M., (1992). The Mediterranean flora and vegetation. Edition: Ouest-France-Rennes. Paris. 221p.
[25] - Iserin. P, Masson. M et Restellini. J.P.,(2001) Encyclopedie des plantes medicinales : Identification, Preparations, Soins, Larousse Paris 2001, PP 10- 14
[26] -Imdorf . A., Bogdanov. S., Ibanez Ochoa. R and Calderone. N., (1999). Use of essential oils for the control of Varroa destructor (Oud.) in honey bee colonies. Apidologie PP209 - 228.
[27]-Jean M. ,(1992). Nouveau guide de phytotherapie remede d'hier pour medecine de demain. Edition Jacque Grancher.
[28]- Jan Vet Jiri S., (1984). Medicinal plants.
[29]-Leon Raul H. (2005). Substitution of synthetic solvents and active materials by a "solvent/active" combination of plant origin. Ecole Nationale Superieure des Ingenieurs en Arts France. PP 18-20.
[30]-Marouf. A., (2000). Dictionary of botany. Edition: dounod paris. PP129-200.
[31]-Nicolas .V., (1991). Essential oil: World production, ëinternational trade and ëprice evaluation. io[eme] journee nationale des huiles essentielles. Proceedings, Ravista italiana. Eppos .PP 534-539.
[32] - Padrini. F and Luchroni .N . , (1996). Le grand livre des huiles essentielles, medecine douce, bien etre. Edition de VECCHIS-A. 206p.
[33] - Paris. R and Moyse. H., (1976) Matiere medicinale. Tome 3. Ed Masson et Cite. Paris. 509p.
[34]- Paris R and Dillemann. G., (1960). Les plantes medicinales des régions arides considerees surtout du point de vue pharmacologique. Laboratoire de matiere medicale, Faculte de pharmacie de Paris.99p.
[35]- Oussalah .M., Caillet .S., Saucier. L. and Lacroix. M., (2006). Antimicrobial activity of selected plant essential oils on the growth of a *Pseudomonas putida* strain isolated from meat. Meat Science PP 236-244.
[36]-Quatresons. N., (2011).L'aspergillose humaine .Diplome d'etat de docteur en pharmacie. P2.
[37]- Saadou. Etude et caracterisation chimique des huiles essentielles du genre Pinus, dans le Parc National d'El kala (P.N.E.K). Memoire de Magister. Edition universite d'Annaba. PP 17-22.
[38]- Sadok. G. ,(2009). La phytotherapie. Ecole superieure des sciences et techniques de la sante de Sousse. P07.
[39] -Sebai M and Boudali M. , (2012). La phytotherapie entre la confiance et mefiance. Institut de formation paramedical CHETTIA. PP 09-21.

[40]- Stahl et al, (1975) Chromatographic and microscopic analysis of drugs. Ed Lavoisier. 380p.
[41]- Suppakul .P., Miltz .J., Sonneveld. K et Bigger .S., (2003). Antimicrobial properties of Basil and its possible application in food packaging. J.Agric. Food Chem, PP 3197-3207
[42]- Vecchi .S., (1993). Encylopedia des plantes medicinales with 200 illustrations. Edition: Tec et DOC. PP7-8.
[43]- Zaidi .N and Hacini . N. ,(2002). Etude ethnobotanique des deux familles : Laliacees et Solanacees dans l'est algerien. Memoire de DES. Universite d'Annaba. P35.
[44]-Zhiri. A and Baudoux. D., (2008). Chemotyped essential oils and their synergies.

WEB SITES

[45] - http://www.cbnbrest.fr/site/pdf/ambroisie.pdf
[46]http://www.franceagrimer.fi7content/download/13937/100977/file/Livret%20Les%20pla
n
tes%20aromatiques%202012.pdf
[47] -http://www.egk.ch/fr/doc/nl/thymian.pdf
[48] -www.phytomania.com/thym.htm
[49]-file:///H:/thym/thym_fichiers/thy
[50] -file:///H:/armoise/index.php fichiers/artemisia.jpg
[51]-http://www2.ac-lyon.fr/enseigne/biotech/microbio/milieux.html

Appendix

Appendix 01:

Sabouraud midfield:

Formula in g/L of distilled water :

- Peptone10 g
- Glucose massë20 g
- Agar-agar15g
- Distilled water 1000 ml
- Vitamins and growth factors
- PH = 6.0 [51]

Appendix 02 :

Composition of the Mueller-Hinton midfield:

In grams per litre of distilled water

- Beef infusion300ml .
- Casëine peptone17 .5
- Starch mat's1 ,5
- Agar17
- Final PH = 7.4 [51]

Presented by : **Ben Adouda Saloua** **and** **Chaib Nadia**	**Evaluation of the antifungal power of *Artemisia* herba alba and *Thymus vulgaris* against several species of human pathogenic fungi (Candida sp and Aspergillus sp).**	Framed by: **DR-BERROUK-Houda**

ReSume:

Thyme (*Thymus vulgarus)* and mugwort (*Artimisia herba alba)* are medicinal plants used in traditional medicine. Hydrodistillation is the main method used to obtain essential oils from these two medicinal plants.
Thin layer chromatography (TLC) shows and identifies the presence of two chemical groups for each oil.
Based on our results, we can confirm that the two essential oils studied, in particular thyme essential oil, are effective natural antifungal agents and can be a very important tool in the fight against fungal diseases.

Keywords : hydro distillation, essential oils, thin layer chromatography, antifungal agents

MIX
Papier aus verantwortungsvollen Quellen
Paper from responsible sources
FSC® C105338

Printed by Books on Demand GmbH, Norderstedt / Germany